advances in genomic
sequence analysis and
pattern discovery

SCIENCE, ENGINEERING, AND BIOLOGY INFORMATICS

Series Editor: Jason T. L. Wang
(New Jersey Institute of Technology, USA)

---

advances in genomic sequence analysis and pattern discovery

editors

Laura Elnitski
National Human Genome Research Institute,
National Institutes of Health, USA

Helen Piontkivska
Kent State University, USA

Lonnie R Welch
Ohio University, USA

World Scientific

NEW JERSEY · LONDON · SINGAPORE · BEIJING · SHANGHAI · HONG KONG · TAIPEI · CHENNAI

Published by

World Scientific Publishing Co. Pte. Ltd.

5 Toh Tuck Link, Singapore 596224

USA office: 27 Warren Street, Suite 401-402, Hackensack, NJ 07601

UK office: 57 Shelton Street, Covent Garden, London WC2H 9HE

British Library Cataloguing-in-Publication Data
A catalogue record for this book is available from the British Library.

ISBN-13 978-981-4327-72-5
ISBN-10 981-4327-72-7

Typeset by Stallion Press
Email: enquiries@stallionpress.com

Printed in Singapore.

Contents

v

Preface

Those who are involved with mapping the genomic landscapes are participating in one of the most exciting frontiers of science. We have the opportunity to reverse engineer the blueprints and the control systems of living organisms. Computational tools are key enablers in the deciphering process. Thus, this book provides an in-depth presentation of some of the important computational biology approaches to genomic sequence analysis. The first part of the book discusses methods for discovering patterns in DNA and RNA. This is followed by the second part that reflects on methods in various ways, including performance, usage and paradigms.

Part I, *Pattern Discovery Methods*, provides a collection of computational methods and tools. Chapter 1, "*Large-Scale Gene Regulatory Motif Discovery with NestedMICA*," presents an algorithmic approach, describes usage of the tool based on the algorithm, and illustrates its usage via a detailed case study. In Chapter 2, "*R'MES: A Tool to Find Motifs with a Significantly Unexpected Frequency in Biological Sequences*," the authors describe a software tool that contains rigorous statistical models of DNA words. "*An Intricate Mosaic of Genomic Patterns at Mid-range Scale*," Chapter 3 of the book, focuses on intricate mosaics found in genomes; a number of specific patterns are identified. The fourth chapter, "*Motif Finding from Chips to ChIPs*," provides a comprehensive survey of methods for the *de novo* discovery of putative over-represented transcription factor binding sites in nucleotide sequences. Part I concludes with a chapter that considers the discovery of RNA structural motifs: "*A New Approach to the Discovery of RNA Structural Elements in the Human Genome*."

The second part, *Performance and Paradigms*, consists of chapters that contemplate the effectiveness of relevant computational biology techniques. Chapter 6, "*Benchmarking of Methods for Motif Discovery in DNA*," presents a variety of metrics for assessing the performance of the class of methods described in Part I. In "*Encyclopedias of DNA Elements for Plant Genomes*," the application of methods is illustrated with case studies. The topic of scalable algorithmic approaches is considered in

Chapter 8, "*Manycore High-Performance Computing in Bioinformatics.*"
Chapter 9, "*Natural Selection and the Genome,*" discusses evolution of
genomic sequences and the role that natural selection plays in directing genome evolution. It also provides a conceptual framework for better
understanding of the evolutionary implications and insights that are generated through genomic sequence analyses, and emphasizes the critical
role of purifying selection.

About the Editors

 Dr. Laura Elnitski is a molecular and computational biologist who studies noncoding functional elements in vertebrate genomes. She has served as an analyst for the Mouse, Rat, Chicken and Bovine Genome Consortia.

Dr. Elnitski is extensively involved in NHGRI's ENCODE (Encyclopedia of DNA Elements) project, which aims to produce a comprehensive catalog of functional elements in the human genome. Dr. Elnitski's research uses integrative analyses to elucidate both the presence and activity of functional elements in the human genome that have been historically difficult to characterize. For example, computationally, her work predicts mutations in coding sequences that affect proper splicing. Targets of these mutations include exonic splicing enhancers and silencers. In experimental analyses, she is mapping elements that silence gene expression using an assay system designed in her lab.

Driving towards a molecular understanding of ovarian cancer, Dr. Elnitski combines *in silico* and wet-bench techniques. She has annotated bidirectional promoters in the human genome, including those regulating noncoding genes, using data collected in RNA-seq assays. These results are being used to find gene silencing events caused by aberrant methylation in tumor samples. She is also addressing functional consequences of mutations in those tumors.

Dr. Elnitski is the recipient of a Ruth L. Kirschstein Postdoctoral Fellowship through the NIH (2000–2003), Outstanding Research Achievement Award (International Symposium on Bioinformatics Research and Applications — 2007), a featured scientist in the *Women in Bioinformatics Research* documentary (2007) and a Genome Technology Young Investigator Award (2009). She serves as an *ad hoc* reviewer for the NIH GCAT Scientific Grant Review Panel and is an associate editor of *BMC Genomics* and formerly *Genome Research*.

She is currently the Head of Genomic Functional Analysis Section of the Genome Technology Branch at the National Human Genome Research Institute, NIH, USA.

Dr. Helen Piontkivska is currently an Assistant Professor in the Department of Biological Sciences at Kent State University. She received her Ph.D. in Genetics from the Pennsylvania State University, and has over 10 years of experience in bioinformatics and evolutionary genomics and published over 35 peer-reviewed publications. Dr. Piontkivska's research is focused on understanding the mechanisms of genome evolution using state-of-the-art bioinformatics approaches, in particular, delineating evolutionary mechanisms responsible for genomic changes in infectious agents, such as viruses, and disease-related genes, including cancer and immune and inflammatory genes using bioinformatics, machine learning and molecular evolutionary and phylogenetic approaches. Her work is funded by the National Institutes of Health. She currently serves on the editorial board of *Molecular Biology and Evolution* and is a member of the steering committee of The Ohio Bioinformatics Consortium.

Professor Lonnie R. Welch received a Ph.D. in Computer and Information Science from the Ohio State University. Currently, he is the Stuckey Professor of Electrical Engineering and Computer Science at Ohio University, and he is a member of the Graduate Faculties of the Biomedical Engineering Program and of the Molecular and Cellular Biology Program. Dr. Welch performs research in the areas of bioinformatics, computational regulatory genomics, machine learning and high performance computing. His research has been sponsored by the National Human Genome Research Institute, the Ohio Plant Biotechnology Consortium, NASA, the National Science Foundation, the Defense Advanced Research Projects Agency, and the Ohio Board of Regents. Dr. Welch has more than 20 years of research experience in the area of high performance computing. In his graduate work at Ohio State

University, he developed high performance 3-D graphics rendering algorithms, and he invented a parallel virtual machine for object-oriented software. For 15 years, his research focused on middleware and optimization algorithms for high performance computing; this work produced three successive generations of adaptive resource management middleware for high performance real-time systems, and resulted in two patents and more than 150 publications. Currently, Professor Welch directs the Bioinformatics Laboratory at Ohio University, where he performs research in the area of computational regulatory genomics.

Dr. Welch is the founder and Chair of the Ohio Collaborative Conference on Bioinformatics and the Great Lakes Bioinformatics Conference (an official conference of the International Society on Computational Biology). As Founding Chair of the Ohio Bioinformatics Consortium, an Affiliated Regional Group of the International Society on Computational Biology, Dr. Welch has been an active member of the Regional Affiliates Committee of the ISCB. He is the Principal Investigator of the $9M Bioinformatics Program which is funded by the Ohio Board of Regents and 11 academic institutions from Ohio. Prof. Welch is founder and Co-Chair (2010, 2011) of the ISMB Special Interest Group on Bioinformatics for Regulatory Genomics (BioRegSIG). He has served on the organizing committees of the 2009 Bioinformatics Open Source Conference, the 2008 ISMB Special Interest Group on Genome-scale Pattern Analysis in the Post-ENCODE Era, the International Symposium on Bioinformatics Research and Applications, and the IEEE International Symposium on Bioinformatics and Bioengineering.

Part I

PATTERN DISCOVERY METHODS

Chapter 1

Large-Scale Gene Regulatory Motif Discovery with NestedMICA

Matias Piipari*, Thomas A. Down†
and Tim J. P. Hubbard*

In this chapter we describe practical uses of computational regulatory motif discovery with the NestedMICA suite, with a focus on higher eukaryotic genomes. The NestedMICA algorithm is contrasted with other popular motif inference approaches. Recent developments of the algorithm and additions to analysis tools included with the suite are also described.

The practical use of NestedMICA is demonstrated with a case study where 120 *Drosophila melanogaster* regulatory motifs are inferred and validated computationally and experimentally. This work was previously published (Down *et al.*, 2007) and is summarized here to demonstrate key principles of genome-scale regulatory motif studies for the research community. The importance of the interplay between computational and experimental work in finding and understanding functional elements is discussed. It is demonstrated how computationally discovered motifs can be associated with several independent lines of supporting evidence for their function, such as tissue-specific gene expression and inter-species conservation pattern. Similarity comparisons between computationally discovered and experimentally verified motif sets are also shown.

A practical tutorial of the NestedMICA is included to introduce the suite to researchers new to such tools. NestedMICA is used to reproduce the binding site motif of the STAT1 transcription factor using ChIP-seq data from a previously published study. We describe several new additions to the suite: a tool for analyzing motif hit overrepresentation in a positive sequence set versus a negative sequence set, a tool for finding statistically significant sequence motif matches, and a series of utilities for retrieving and preprocessing sequence regions for motif inference experiments.

---

*Wellcome Trust Sanger Institute, Hinxton, Cambridgeshire, UK
†Wellcome Trust/Cancer Research UK Gurdon Institute, Cambridge, Cambridgeshire, UK

1. Introduction

Computational discovery of transcription factor (TF) binding sites using short motif discovery algorithms from genomic DNA sequences is a long-standing research problem. It has motivated computational biologists to propose literally hundreds of algorithms over the course of over 30 years, beginning from a pioneering paper in 1977 where pairwise comparisons of aligned sequence immediately close to prokaryotic transcription start sites (TSS) and terminator sequences were used to infer recurring motifs (Korn *et al.*, 1977).

Binding site motif inference methods on their own have proven far from perfect in predicting functional TF binding sites in higher eukaryotic genomes. Transcriptional regulation is now known to be a complex landscape of regulatory signals not only encoded in short (5–20 bp) transcription binding site motifs proximal to transcription start sites. Distant-acting enhancers (Visel *et al.*, 2009), chromatin state (Cairns, 2009) and epigenetic signals (Jaenisch and Bird, 2003) are also thought to contribute to gene regulation. Promoters in higher eukaryotes are also large and variable in size, and alternative transcripts are common (Zhu and Halfon, 2009). It should therefore not come as a surprise that inferring transcription factor binding site motifs, let alone functional sites, remains difficult. As we will show, development and application of such methods is still highly important. Recent reviews (Das and Dai 2007; MacIsaac and Fraenkel, 2006; Nguyen and Androulakis, 2009; Sandve and Drabløs, 2006) considered and categorized subsets of motif inference methods according to varying criteria. We discuss here only the most essential factors to consider for a researcher new to the field: motif model used, the types of data that are considered in context of the genomic sequence, and the scalability of the methods for genomic scale analysis of regulatory sequences.

The common principle in regulatory motif discovery studies is the use of computational methods to infer enriched sequence word signals from non-coding sequences associated with a set of genes. Input sequences of fixed length for motif discovery are usually taken from around the TSS of a gene set of interest. More specific regions of interest identified through upstream analyses can also be used when supporting evidence is available (e.g. ChIP-chip or ChIP-seq derived "peak" regions). It has only recently become possible to computationally infer large numbers of regulatory motifs from a considerable fraction of genes of a whole eukaryotic organism in a single large experiment (Down *et al.*, 2007; Xie *et al.*, 2005).

1.1. *Assessment of motif inference tools*

The motif hit finding sensitivity of some of the most commonly used *ab initio* motif discovery methods has been benchmarked by Tompa *et al.* (2005). *Ab initio* refers to methods that consider sequence alone in search of recurring signals. Binding site sequences from the TRANS-FAC transcriptional regulation database (Matys *et al.*, 2006) were "spiked" to sequences that were either synthetic or originated from randomly chosen promoter sequence (assumed to be devoid of motifs other than those inserted). The overlap of match positions of the best-ranking motif inferred by each of the benchmarked tools was then used in the assessment. It should be noted that a standard methodology or statistical criteria for assessing regulatory motif inference performance have not surfaced in the research community. This is not surprising given the variety of aims of the different tools and scales of dataset sizes at which each method operates on, and more pressingly the low number of "gold standard" regulatory regions that have been annotated in depth. We still lack the necessary understanding of site-specific gene regulatory signals to accurately call sequence specific gene regulatory binding events on the needed scale. Large-scale transcription factor ChIP-sequencing experiments conducted as part of the ENCODE scale-up project hold great potential for providing such benchmarking data for computational motif discovery tools.

1.2. *What is a motif?*

Arguably the most important distinction between motif discovery tools is the model that is used as a description of the inferred sequence signals. Oligonucleotide word enumeration methods are the classical approach that is still used in several recent algorithms, ranging from reporting ranked k−mers of a specified length (van Dongen *et al.*, 2008) to IUPAC consensus strings that allow for describing degeneracy in positions (Marschall and Rahmann, 2009) to regular expression like patterns which also allow degenerate and gap positions (Xie *et al.*, 2005). Enumeration-based methods are generally computationally light and therefore very fast, but are not

E47 (M00071)

| 1 | 2 | 3 | 4 | 5 | 6 | 7 | 8 | 9 | 10 | 11 | 12 |

Fig. 1. A sequence logo depiction of the 12-column long human E47 (TAL1) transcription factor motif (Hsu *et al.*, 1994) deposited in the TRANSFAC regulatory motif database (Matys *et al.*, 2006). Relative height of each nucleotide in a column corresponds to its weight in the probability distribution for the position. Columns are scaled by information content to depict the strength of the bias of nucleotides at that position.

capable of modeling long TFBS patterns, or those with a large number of weakly constrained positions. Probabilistic models such as position weight matrix (PWM) preserve more of the information of individual motif positions (columns) and PWMs have been shown to systematically perform better in describing regulatory binding site patterns (Osada *et al.*, 2004). The PWM, introduced by Stormo *et al.* (1982) to motif discovery, consists of independent column-wise probability distributions. They are often visualized with the "sequence logo" representation (Fig. 1) introduced by Schneider and Stephens (1990).

The PWM is well suited for defining sequence motifs that have a mixture of strongly and weakly constrained positions. It provides a more accurate and readily interpretable description of binding specificity of most transcription factors than oligonucleotide word based models. It however also lacks the capacity to describe possible interdependencies or variable length linker positions in binding site motif, both known effects for certain transcription factors (Benos *et al.*, 2002). A recent protein-binding microarray-based *in vitro* profiling of 104 different mouse transcription factor motifs (Badis *et al.*, 2009) suggests that "in between motif position" variation in binding specificity is widespread amongst higher eukaryotic transcription factors. Badis *et al.*, did not report the number of factors studied for which multiple PWM models are needed for describing high-affinity binding sites (additional PWMs per TF are reported to be necessary mostly for describing lower affinity sites). The scarcity of *in vivo* binding site specificity data makes it premature to assess whether more complex probabilistic models than PWM are generally necessary for accurately recording TF specificity patterns. More complex motif models based on Bayesian networks (Barash *et al.*, 2003) or Markov networks (Sharon *et al.*, 2008) that have been developed to address the shortcomings of the PWM tend to suffer from issues associated with parameter estimation from the available data: it is hard to estimate more complex motif models that generalize to sequences other than those used for model training (problem known as "overfitting"). PWM therefore remains a central model in computational motif discovery.

1.3. *Motif inference with additional supporting data*

Some motif inference algorithms consider other data in addition to genomic sequence to support the inference task. Especially gene expression data have been made use of in various approaches. The earliest method for this is gene expression clustering based division of genes to sets followed by motif inference from the individual sets (Roth *et al.*, 1998). More sophisticated methods have been developed for gene expression, beginning from the pioneering regression-based framework REDUCE by Bussemaker *et al.* (2001) that sparked publication of several related methods (Conlon *et al.*, 2003; Foat *et al.*, 2005; Kechris and Li, 2008; Keleş *et al.*, 2002). Ranking or weighting sequences based on ChIP-chip experiments has also been applied in motif discovery (Conlon *et al.*, 2003; Liu *et al.*, 2002).

Phylogenetic foot printing methods that apply sequence conservation, alignment of orthologous sequences and phylogenetic models of regulatory regions to improve sensitivity in detecting regulatory motif and *cis*-regulatory modules also show great promise (Siddharthan, 2008). In addition to sequence conservation, other nucleotide position specific properties have been connected to TF binding potential in motif discovery algorithms, including for example predicted nucleosome occupancy (Narlikar *et al.*, 2007) and DNA duplex stability (Gordân and Hartemink, 2008).

1.4. *The NestedMICA algorithm*

NestedMICA is a probabilistic *ab initio* motif discovery algorithm that can be used to discover motifs in either DNA (Down *et al.*, 2007; Down and Hubbard, 2005) or protein sequence (Dogruel *et al.*, 2008). Like many other probabilistic algorithms for the purpose, it treats the input sequence as a combination of points of interest (motifs) and background sequence that is generated by a random process. We call this model consisting of a series of motifs and a background a "Sequence Mixture Model" or SMM

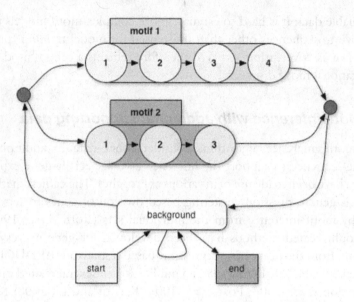

Fig. 2. The default sequence mixture model used in NestedMICA. Numbers inside the motif model nodes denote different columns of the motif. A part of the motif inference problem is optimizing the parameters for a SMM as shown above: the number of columns used for each of the motifs, and the parameters of each of the probability distributions.

(the simplest SMM used in NestedMICA is described as a hidden Markov model in Fig. 2).

NestedMICA uses an independent component analysis (ICA) based motif discovery algorithm that simultaneously optimizes a model for a set of motifs and their "occupancy" (which motif is found from which sequence, depicted in Fig. 3(a)). The motif set model has five key parameters: the number of motifs to estimate, the minimum and maximum length of motifs, the fraction of sequences expected to contain at least one hit to each of the sequence motifs ("usage fraction"), and a sequence background model.

The particular SMM used in NestedMICA addresses several shortcomings of common sequence models used in motif discovery algorithms. Firstly, homotypic and heterotypic clustering of TFBSs that is known to occur commonly in both promoter and enhancer elements (Arnone and Davidson, 1997) is modeled naturally by the NestedMICA SMM that allows for simultaneous modeling of multiple matches of regulatory motifs

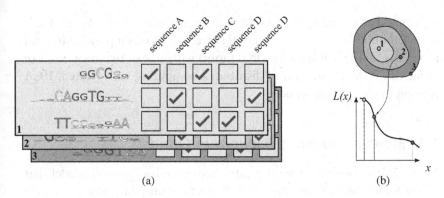

(a) (b)

Fig. 3. Schematic depiction of nested sampling. (a) At each step of the algorithm, there exists a fixed size set of mixing matrices consisting of a set of motifs and their occupancy. The set is ranked by likelihood and is evolved with Monte Carlo moves. (b) The lowest-ranking member of the ensemble is removed at each iteration and a new sample is drawn from the prior until a solution with a higher likelihood is found.

per sequence. This model improves the sensitivity and runtime of the method in especially large motif inference tasks by making it unnecessary to repeat a run of the algorithm after masking motif instances. Repeated runs are needed for example by the popular MEME (Bailey *et al.*, 2006). In NestedMICA, the set of motifs and their occurrence are estimated simultaneously as a "mixing matrix". Both the motif column parameters and their occurrence are optimized within the same iterative sampling process (Fig. 3(a)). A specific SMM is included for inferring specifically motifs that show heteroclustering.

1.5. *Nested sampling*

NestedMICA uses a Markov Chain Monte Carlo (MCMC) inference strategy called nested sampling which does not require heuristics to provide suitable local starting points for motif search, or require repeated restarts of the algorithm like Gibbs samplers (Lawrence *et al.*, 1993) and greedy expectation maximization based procedures such as that applied in the popular MEME (Bailey *et al.*, 2006). A schematic of the NestedMICA nested sampling procedure is provided in Fig. 3.

Scalability and runtime performance were a key consideration in the design and implementation of NestedMICA. The tool makes use of multiple CPUs when available and the computational load can also be distributed over multiple computers. The algorithm scales to 40 CPU cores (unpublished data). In practical terms, this makes it possible to train motif sets on the scale of a hundred motifs with at least two megabases of sequence. A demonstration of distributed computing with NestedMICA is given in the tutorial section.

1.6. *Mosaic sequence background model*

NestedMICA provides a sophisticated sequence background model that allows for modeling compositionally distinct regions, for example the variation in GC content that is known to occur on multiple scales in eukaryotic genomes (Burge *et al.*, 1992; FitzGerald *et al.*, 2006). We call this background model "mosaic" to highlight its capability to describe sequence as a mixture of multiple generative processes (Markov chains). Use of multiple Markov chains, or "classes", that are weighted per sequence position improves the capacity of the background to describe compositional biases and is considerably a less complex model than a high-order Markov chain background that is commonly used in motif inference algorithms. Interestingly, a recently published motif inference algorithm BayesMD which uses a background model approach related to that of NestedMICA improves sensitivity over MEME, Align-ACE, MDScan, and also against NestedMICA in most benchmarks (Tang *et al.*, 2008).

2. Results

Since its publication, NestedMICA has been used in a number of regulatory genomics studies of both human and other organisms. Examples include analysis of Polycomb and Trithorax binding sites in *Drosophila* (Kwong *et al.*, 2008), zebrafish distal enhancers (Rastegar *et al.*, 2008), targets of the transcription factor Ntl (Morley *et al.*, 2009), indirect targets of the deafness associated micro-RNA miRNA-96 in mouse (Lewis *et al.*, 2009), as well as transcription factors involved in determination of ES cell transcriptional programs in mouse (Chen *et al.*, 2008; Loh *et al.*, 2006). Here, we will concentrate on work done by Down *et al.* (2007) as a case study of applying NestedMICA for genome scale motif discovery, namely the discovery of 120 enriched motifs from proximal *Drosophila* promoters. The analysis of this motif set predicted from putative promoters of over 2400 *Drosophila* genes provides validation of genome scale *ab initio* motif inference in discovering gene regulatory patterns from eukaryotic genomes. This study demonstrates certain principles that are important in such computational studies: over-represented motifs should be rigorously analyzed in the context of other biological data such as gene expression, conservation and experimentally determined motif sets.

2.1. Choice of sequence regions for motif inference

The first key step in genomic DNA motif discovery is the choice of search regions. In Down *et al.* (2007), a conscious choice was made to not constrain input sequences by conservation. Sequence conservation of the predicted motifs was applied as an independent validation of the predicted motifs. The motifs were trained from 200-base long sequences from positions −200 to −1 of each of the 2424 annotated genes from the *D. melanogaster* chromosome arm 2L with overlaps and mono- and di-nucleotide repeats removed. This yields over 422 kb of sequence. The benefits of weighting a motif inference experiment by inter-species conservation to directly aim the experiment is not clear; highly conserved noncoding sequences are known to contain regulatory elements, especially those that are developmentally active (Visel *et al.*, 2008), but binding site turnover and by this mechanism "shifting" of functional TFBSs between related species has been observed in both mammalian (Dermitzakis and Clark, 2002) and more closely related *Drosophila* regulatory sequences (Costas *et al.*, 2003; Emberly *et al.*, 2003). It is also worth remembering that a conservation score contains an implicit assumption of a correct alignment between different organisms, which is hard to guarantee for noncoding sequences of higher eukaryotic genomes.

One hundred and twenty motifs of 12-nt length were trained from the above-mentioned input sequences in a single experiment, using a background model trained from the same sequences. Notably the version 0.7 of NestedMICA used in this work required a fixed length motif parameter — motifs were retrained from initial motif match sequences by choosing the motif length that gave optimal Bayesian evidence. The more recent NestedMICA version 0.8 optimizes the motif length as part of the sampling procedure and additional post-processing of the results is therefore no longer required.

2.2. Finding significant matches

After motif discovery, a score cutoff was calculated for each of the motifs as a prerequisite to determining the statistically significant motif match positions in the *Drosophila* genome. The cutoffs were assigned using a binomial test ($p < 0.05$). The motif score cutoff assignment protocol is described in more detail by Down *et al.* (2007), but in simple terms it is used to find the point in the motif hit score distribution at which random samples from the background model produce scores that resemble the observed distribution (as measured by a binomial test). This is in contrast to most motif discovery studies where a binding site p-value cutoff is derived assuming a background model parameterized essentially by nucleotide content.

2.3. Comparison of NestedMICA Drosophila motifs against reference motifs

The NestedMICA *Drosophila* motifs were compared with several reference motif sets to find reciprocally matching motif pairs. The reference motifs included a 10-motif set from an earlier computational study of *Drosophila* core promoters (Ohler *et al.*, 2002), 15 consensus motifs derived from positionally biased octamers (FitzGerald *et al.*, 2006), 30 motifs trained from DNAse I footprinting derived binding site data from the FlyReg database (Bergman *et al.*, 2005), as well as 172 motifs containing the JASPAR database (Sandelin *et al.*, 2004) extended with additional 49 SELEX- or consensus-based *Drosophila* motifs from primary publications referenced in the FlyReg database. Match between NestedMICA motifs and each of the reference sets was measured with an Euclidian-like distance metric (Down *et al.*, 2007) in a reciprocal match comparison similar to the procedure used to identify orthologous genes from genomes. An empirical p-value was calculated for each identified reciprocal match using a distribution of hit scores between random reciprocal match between the target motifs with shuffled reference motifs. A total of 25 significant matches ($p < 0.05$) were identified, suggesting that NestedMICA is able to capture regulatory motifs that closely match previously described ones, but also discovers a number of novel ones. Overlap of sequence-level matches between NestedMICA motifs and the corresponding reference motifs was determined. Interestingly, it

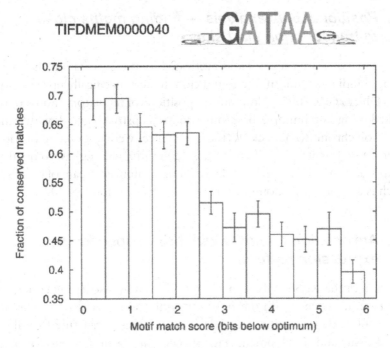

Fig. 4. High motif match score is associated with a high fraction of conserved matches. Adapted from Fig. 5 in Down *et al.*, 2007 (doi:10.1371/journal. pcbi.0030007.g005).

was found that although there is overlap in bases covered, for almost half of the motifs the overlap in matches falls to a range of 10–50% of bases covered.

2.4. Sequence conservation analysis of motif matches

Conservation of motif hits was assessed to identify motifs that show evidence for functional constraint. Non–protein-coding sequence alignments of the motif match positions between two closely related *Drosophila* species *D. simulans* and *D. yakuba* (excluding those sequences used for Nested-MICA motif discovery) were assessed. The majority of motifs (78/120) show a statistically significant correlation between motif score and degree of conservation (example shown in Fig. 4).

2.5. Positional bias analysis — finding motifs close to transcription start sites

The choice of search regions was motivated by the expectation that regulatory binding site motifs are found close to and specifically upstream of TSS (Ohler *et al.*, 2002). Nonuniform positions of regulatory motifs were studied using 200 hundred base long windows upstream and downstream of TSS on chromosome arm 2R (these sequences were not used in the discovery experiment). For 70 out of the 120 motifs, the highest frequency of hits was found to be within 400 bases immediately upstream of the TSS, which is a significantly nonuniform position distribution ($p < 0.01$).

2.6. Association of motifs with tissue-specific gene expression pattern

Function of the NestedMICA motifs was studied by associating core promoter motif matches with the *in situ* hybridization dataset of Tomancak *et al.* (2002) that contains alongside *in situ* images a structured vocabulary of tissue and development stage specific expression patterns for over 2000 *D. melanogaster* genes (the ImaGO ontology). All 120 motifs were scanned against the *D. melanogaster* genes, using score thresholds defined above. The number of times that each motif was found to be associated with an expression pattern term was recorded (either directly with a term or through the ImaGO ontology hierarchy). Motif labels of individual hit records were shuffled to determine empirical p-values for the observed motif-expression associations. To account for multiple testing (motif tested against all the expression terms), motif identifiers were shuffled prior to repeating the p-value calculations and the distribution of shuffled p-values was compared against the correctly labeled p-values. This comparison shows that many motifs associate strongly with at least one ImaGO term (Fig. 5). Expression-associated motifs include several previously uncharacterized motifs. This suggests that large fraction of the discovered motifs is indeed involved in time- or tissue-specific transcriptional regulation.

An example of a previously known association identified by NestedMICA is given by Down *et al.* (2007) in the form of an *srp* like motif whose associated expression patterns (developing fat body and amnioserosa) match those described in prior literature for *srp*. This example also shows a weakness of motif inference and expression analysis alone in mapping

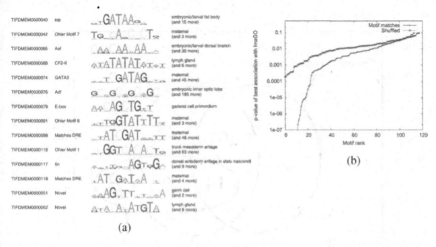

(a)

(b)

Fig. 5. (a) Examples of the 25 NestedMICA motifs that show association with ImaGO embryonic expression ontology terms. (b) Significance scoring of shuffled motif labels indicate the false discovery rate at different p-value thresholds. Adapted from Down *et al.*, 2007 (doi:10.1371/journal.pcbi.0030007.g007).

regulatory motifs to gene regulatory events: *srp* has five paralogs in the *Drosophila* genome, including *pnr* that has a similar sequence specificity pattern (Haenlin *et al.*, 1997), and is also known to be expressed and required for the survival of the developing amnioserosa (Ramain *et al.*, 1993; Winick *et al.*, 1993).

This study demonstrates methods for and importance of analyzing computationally inferred regulatory motifs in the context of different lines of evidence. Although a relatively small fraction of the motifs is supported in a statistically significant way by any of the analyses described, combining their results shows that the great majority of the motifs have a significant association with either a specific expression pattern, sequence conservation or positional bias (Fig. 6).

3. Motif Inference Tutorial

In this chapter, we have described a study where the NestedMICA suite was applied to a genome-scale computational regulatory motif inference task. However, we have recently developed an additional suite of sequence retrieval and motif set analysis tools to complement NestedMICA. We call this set of tools NMICA-extra. To demonstrate the application of

All motifs (n=120)

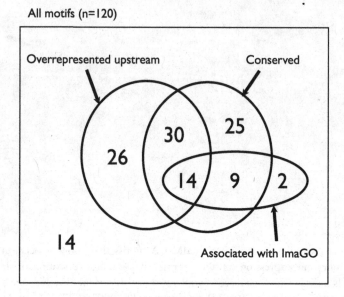

Fig. 6. Summary of lines of evidence supporting the 120 motifs discovered in Down *et al.*, 2007 (doi: 10.1371/journal.pcbi.0030007.g009).

NestedMICA and NMICA-extra to the increasingly important task of motif inference from ChIP-seq data, we will use these tools to recreate the STAT1 transcription factor binding motif described by Robertson *et al.* (2007) with the ChIP-seq peak region data from Robertson *et al.* (2008). A more detailed version of the tutorial with command line examples and installation instructions for the necessary software is available at http://wiki.github.com/mz2/nmica-extra.

3.1. *Sequence retrieval and preprocessing*

The first step in the exercise is retrieving input genomic sequences corresponding to the ChIP-seq peak regions. To ease the retrieval and preprocessing of input sequence (repeat masking and exclusion of translated sequences), NMICA-extra includes tools for retrieving sequence from the Ensembl database (Flicek *et al.*, 2008), namely **nmensemblseq**, **nmensemblfeat** and **nmensemblpeakseq**. The first of the three, **nmensemblseq**, can be used to retrieve sequences around transcription start sites or 3′ UTRs or introns. **nmensemblfeat** is used for retrieving specific sequence regions using GFF formatted annotation files as input.

We use **nmensemblpeakseq** to retrieve sequence windows corresponding to 50-base-long sequence windows around ranked ChIP-sequencing peak maximum positions of the 500 top-ranking peaks. **nmensemblpeakseq** supports several common peak caller formats: MACS (Zhang *et al.*, 2008), FindPeaks (Fejes *et al.*, 2008), SWEMBL, as well as the more generic BED and GFF annotation data formats.

3.2. Background model estimation

Before conducting motif inference from the retrieved sequences, it is advisable to estimate a sequence background model from the input sequences as a separate step. This can be done with the command **nmmakebg** that requires two input parameters: Markov chain order and the number of mosaic classes. The Markov chain parameter is usually set to 1st order because some of the DNA motif specific downstream analysis tools included in the suite require this. Four mosaic classes tend to yield the best performance with eukaryotic noncoding sequence (Down and Hubbard, 2005). It is however best to evaluate different mosaic class parameters before the potentially long-running motif inference analysis. Background models can be evaluated using the command **nmevaluatebg**.

The output of **nmevaluatebg** can be used to find the mosaic order parameters at which the background model performance, as measured by sequence likelihood given the background model, shows little increase or drops. These parameter values are then taken as the optimal ones.

The evaluation shown in Fig. 7 suggests four as an appropriate value for the mosaic count parameter for our ChIP-seq peak sequences (sequence likelihood shows little improvement with five or more classes). We therefore trained the background model with four classes.

3.3. Motif inference

After retrieving the input sequences and determining class and order parameters with **nmmakebg**, we run the NestedMICA motif inference tool **nminfer** to discover a single motif from the input data. For long running NestedMICA tasks, it is helpful to instruct it to output checkpoint files that can be used to restart the computation, as well as sample motif set files that can be used to visualize the state of the sampling periodically. Please refer to the online version of the tutorial for details.

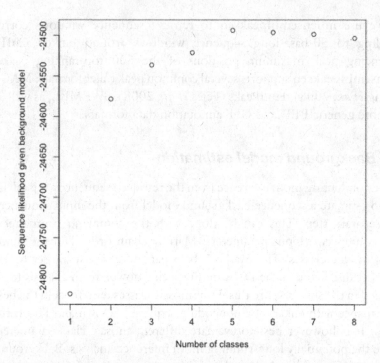

Fig. 7.   Evaluation of different sequence background model class counts with meval-uatebg (1st order Markov chain) with the STAT1 ChIP-seq peak sequences. The sequence likelihood is shown to level after four classes.

3.4. *Visualizing NestedMICA motifs as sequence logos*

NestedMICA's output can be visualized with a number of tools, including the cross-platform **MotifExplorer,**[a] the OS X only **iMotifs** (Piipari *et al.*, 2010) (Fig. 8), or the **seqLogo** package supplied as part of BioConductor (Gentleman *et al.*, 2004).

3.5. *Motif overrepresentation analysis*

When interpreting the output of NestedMICA, it is important to note that the algorithm does not rank its output motifs relative to one another or predict hit positions for them. A common way of assessing computationally

---

[a]http://www.sanger.ac.uk/Software/analysis/nmica/mxt.shtml

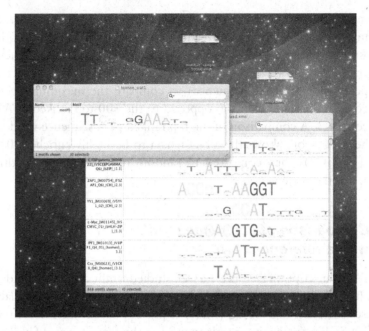

Fig. 8. A screenshot of iMotifs — a motif viewer and analysis tool for OS X that also includes an integrated NestedMICA tool suite.

inferred motifs is by a motif overrepresentation analysis. Overrepresentation analysis is a statistical exercise where sequences with the motif (the positive set) are discriminated from those assumed to be devoid of it (the negative set).

The approach taken in NMICA-extra for calculating the degree of overrepresentation in a set of sequences is the ROC-AUC (Receiver-Operator Characteristic Area Under the Curve) statistic, computed with the tool **nmrocauc**. In short, sequences are labeled as positive or negative and the maximum motif bit score is used to predict if any given sequence is part of the positive or the negative sequence set. The AUC statistic that is reported by this analysis is a measure of how often a randomly chosen positive sequence is ranked above a randomly chosen negative sequence. It therefore provides a measure of separation of maximum motif hit score distribution of the positive examples from the negative examples. To estimate the null distribution of scores with the length distribution and sequence composition used, the negative sequences are shuffled and the randomly generated sequences are then scored according to the same criterion. The shuffling conducted as part of this method accounts for the fact that the maximum hit score distributions of sequences can vary based on nucleotide composition. We retrieve 1000 random core promoter sequence fragments of length 50 bp (in between $-900\,\mathrm{nt}$ and $+100$ relative to TSS, excluding repeats and translated sequence) to compare the maximum scores achieved with these fragments as opposed to the 500 top ranked STAT1 peak sequences. A comparison of the top ranked peak sequence windows with the AUC score shows the STAT1 data to be highly enriched: 0.99 ($p < 1 \times 10^{-5}$) when compared to random noncoding nonrepetitive sequence of the same genome.

3.6. *Comparison of sequence motifs with a reference motif set*

The STAT transcription factors are a well studied family and DNA binding motifs have therefore been recovered and deposited to publicly available databases such as TRANSFAC (Matys *et al.*, 2006) and JASPAR (Bryne *et al.*, 2008). This makes it possible to validate the sequence motif we have inferred from the ChIP-seq data with NestedMICA by searching it against motif databases with the reciprocal matching procedure described above. Reciprocal matching of motifs is implemented in the tool **nmshuffle** that is distributed as part of NestedMICA.

A statistically significant match is identified for the NestedMICA STAT1 motif in the TRANSFAC database ($p < 1 \times 10^{-5}$), and an inspection of the closest matching motifs makes it clear that NestedMICA infers a very similar binding specificity pattern for STAT1 as has been previously reported for members of the STAT family transcription factors (Fig. 9).

4. Conclusions

In this chapter, we outlined the use of *ab initio* motif discovery algorithm NestedMICA in computational discovery of higher eukaryotic transcription factor binding site motifs. We validated the sequence motif signals by associating them with tissue-specific gene expression, positional bias and inter-species conservation patterns. We also show similarity comparisons between computationally discovered and experimentally verified motif sets.

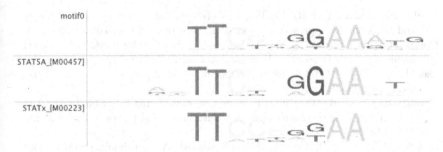

Fig. 9. Comparison of the motif inferred by NestedMICA from the ChIP-seq data (motif0) compared with close hits from the TRANSFAC database.

The NestedMICA case study and tutorial demonstrate in practical terms how a researcher new to regulatory genomics can make use of motif discovery tools to identify and analyze sequence motifs from genomic sequences.

References

Arnone MI, Davidson EH. (1997) The hardwiring of development: Organization and function of genomic regulatory systems. *Development* **124**: 1851–1864.

Badis G, Berger MF, Philippakis AA *et al.* (2009) Diversity and complexity in DNA recognition by transcription factors. *Science* **324**: 1720–1723.

Bailey TL, Williams N, Misleh C, Li WW. (2006) MEME: Discovering and analyzing DNA and protein sequence motifs. *Nucleic Acids Res* **34**: W369–373.

Barash Y, Elidan G, Friedman N, Kaplan T. (2003) Modeling dependencies in protein-DNA binding sites. *Proceedings of the seventh annual international conference ....*

Benos PV, Bulyk ML, Stormo GD. (2002) Additivity in protein-DNA interactions: How good an approximation is it? *Nucleic Acids Res* **30**: 4442–4451.

Bergman CM, Carlson JW, Celniker SE. (2005) Drosophila DNase I footprint database: A systematic genome annotation of transcription factor binding sites in the fruitfly, Drosophila melanogaster. *Bioinformatics* **21**: 1747–1749.

Bryne JC, Valen E, Tang M-HE *et al.* (2008) JASPAR, the open access database of transcription factor-binding profiles: New content and tools in the 2008 update. *Nucleic Acids Res* **36**: D102–106.

Burge C, Campbell AM, Karlin S. (1992) Over- and under-representation of short oligonucleotides in DNA sequences. *Proc Natl Acad Sci USA* **89**: 1358–1362.

Bussemaker HJ, Li H, Siggia ED. (2001) Regulatory element detection using correlation with expression. *Nat Genet* **27**: 167–171.

Cairns BR. (2009) The logic of chromatin architecture and remodelling at promoters. *Nature* **461**: 193–198.

Chen X, Xu H, Yuan P *et al.* (2008) Integration of external signaling pathways with the core transcriptional network in embryonic stem cells. *Cell* **133**: 1106–1117.

Conlon EM, Liu XS, Lieb JD, Liu JS. (2003) Integrating regulatory motif discovery and genome-wide expression analysis. *Proc Natl Acad Sci USA* **100**: 3339–3344.

Costas J, Casares F, Vieira J. (2003) Turnover of binding sites for transcription factors involved in early Drosophila development. *Gene* **310**: 215–220.

Das MK, Dai H-K. (2007) A survey of DNA motif finding algorithms. *BMC Bioinformatics* **8**(Suppl 7): S21.

Dermitzakis ET, Clark AG. (2002) Evolution of transcription factor binding sites in Mammalian gene regulatory regions: Conservation and turnover. *Mol Biol Evol* **19**: 1114–1121.

Dogruel M, Down T, Hubbard T. (2008) NestedMICA as an ab initio protein motif discovery tool. *BMC Bioinformatics* **9**: 19.

Down TA, Bergman CM, Su J, Hubbard TJP. (2007) Large-Scale Discovery of Promoter Motifs in Drosophila melanogaster. *PLoS Comput Biol* **3**: e7.

Down TA, Hubbard TJP. (2005) NestedMICA: Sensitive inference of over-represented motifs in nucleic acid sequence. *Nucleic Acids Res* **33**: 1445–1453.

Emberly E, Rajewsky N, Siggia ED. (2003) Conservation of regulatory elements between two species of Drosophila. *BMC Bioinformatics* **4**: 57.

Fejes AP, Robertson G, Bilenky M *et al.* (2008) FindPeaks 3.1: A tool for identifying areas of enrichment from massively parallel short-read sequencing technology. *Bioinformatics* **24**: 1729–1730.

FitzGerald PC, Sturgill D, Shyakhtenko A *et al.* (2006) Comparative genomics of Drosophila and human core promoters. *Genome Biol* **7**: R53.

Flicek P, Aken BL, Beal Ballester B *et al.* (2008) Ensembl 2008. *Nucleic Acids Res* **36**: D707–714.

Foat BC, Houshmandi SS, Olivas WM, Bussemaker HJ. (2005) Profiling condition-specific, genome-wide regulation of mRNA stability in yeast. *Proc Natl Acad Sci USA* **102**(49): 17675–17680.

Gentleman RC, Carey VJ, Bates DM *et al.* (2004) Bioconductor: Open software development for computational biology and bioinformatics. *Genome Biol* **5**: R80.

Gordān R, Hartemink AJ. (2008) Using DNA duplex stability information for transcription factor binding site discovery. *Pacific Symposium on Biocomputing*, pp. 453–464.

Haenlin M, Cubadda Y, Blondeau F *et al.* (1997) Transcriptional activity of pannier is regulated negatively by heterodimerization of the GATA DNA-binding domain with a cofactor encoded by the u-shaped gene of Drosophila. *Genes Dev* **11**: 3096–3108.

Hsu HL, Huang L, Tsan JT *et al.* (1994) Preferred sequences for DNA recognition by the TAL1 helix-loop-helix proteins. *Mol Cell Biol* **14**: 1256–1265.

Jaenisch R, Bird A, (2003) Epigenetic regulation of gene expression: How the genome integrates intrinsic and environmental signals. *Nat Genet* **33**(Suppl): 245–254.

Kechris K, Li H, (2008) c-REDUCE: Incorporating sequence conservation to detect motifs that correlate with expression. *BMC Bioinformatics* **9**: 506.

Keleş S, van der Laan M, Eisen MB. (2002) Identification of regulatory elements using a feature selection method. *Bioinformatics* **18**(9): 1167–1175.

Korn LJ, Queen CL, Wegman MN. (1977) Computer analysis of nucleic acid regulatory sequences. *Proc Natl Acad Sci USA* **74**: 4401–4405.

Kwong C, Adryan B, Bell I *et al.* (2008) Stability and dynamics of polycomb target sites in Drosophila development. *PLoS Genetics* **4**(9): e1000178.

Lawrence CE, Altschul SF, Boguski MS *et al.* (1993) Detecting subtle sequence signals: A Gibbs sampling strategy for multiple alignment. *Science* **262**: 208–214.

Lewis MA, Quint E, Glazier AM *et al.* (2009) An ENU-induced mutation of miR-96 associated with progressive hearing loss in mice. *Nat Genet* **41**: 614–618.

Liu X, Brutlag D, Liu J. (2002) An algorithm for finding protein — DNA binding sites with applications to chromatin-immunoprecipitation microarray experiments. *Nat Biotechnol* **20**: 835–839.

Loh Y, Wu Q, Chew J *et al.* (2006) The Oct 4 and Nanog transcription network regulates pluripotency in mouse embryonic stem cells. *Nat Genet* **38**: 431–440.

MacIsaac KD, Fraenkel E. (2006) Practical strategies for discovering regulatory DNA sequence motifs. *PLoS Comput Biol* **2**(4): e36.

Marschall T, Rahmann S. (2009) Efficient exact motif discovery. *Bioinformatics* **25**: i356–i364.

Matys V, Kel-Margoulis OV, Fricke E *et al.* (2006) TRANSFAC and its module TRANSCompel: Transcriptional gene regulation in eukaryotes. *Nucleic Acids Res* **34**: D108–D110.

Morley RH, Lachani K, Keefe D *et al.* (2009) A gene regulatory network directed by zebrafish No tail accounts for its roles in mesoderm formation. *Proc Natl Acad Sci USA* **106**: 3829–3834.

Narlikar L, Gordân R, Hartemink AJ (2007) A nucleosome-guided map of transcription factor binding sites in yeast. *PLoS Comput Biol* **3**: e215.

Nguyen TT, Androulakis IP. (2009) Recent advances in the computational discovery of transcription factor binding sites. *Algorithms* **2**: 582–605.

Ohler U, Liao G-C, Niemann H, Rubin GM. (2002) Computational analysis of core promoters in the Drosophila genome. *Genome Biol* **3**: RESEARCH0087.

Osada R, Zaslavsky E, Singh M. (2004) Comparative analysis of methods for representing and searching for transcription factor binding sites. *Bioinformatics* **20**: 3516–3525.

Piipari M, Down TA, Saini H *et al.* (2010) iMotifs: An integrated sequence motif visualization and analysis environment, *Bioinformatics* **26**(6): 843–844.

Ramain P, Heitzler P, Haenlin M, Simpson P. (1993) pannier, a negative regulator of achaete and scute in Drosophila, encodes a zinc finger protein with homology to the vertebrate transcription factor GATA-1. *Development* **119**: 1277–1291.

Rastegar S, Hess I, Dickmeis T *et al.* (2008) The words of the regulatory code are arranged in a variable manner in highly conserved enhancers. *Dev Biol* **318**: 366–377.

Robertson AG, Bilenky M, Tam A *et al.* (2008) Genome-wide relationship between histone H3 lysine 4 mono- and tri-methylation and transcription factor binding. *Genome Res* **18**: 1906–1917.

Robertson G, Hirst M, Bainbridge M *et al.* (2007) Genome-wide profiles of STAT1 DNA association using chromatin immunoprecipitation and massively parallel sequencing. *Nature Methods* **4**: 651–657.

Roth FP, Hughes JD, Estep PW, Church GM. (1998) Finding DNA regulatory motifs within unaligned noncoding sequences clustered by whole-genome mRNA quantitation. *Nat Biotechnol* **16**: 939–945.

Sandelin A, Alkema W, Engstrom P *et al.* (2004) JASPAR: An open-access database for eukaryotic transcription factor binding profiles. *Nucleic Acids Res* **32**: D91–D94.

Sandve GK, Drabløs F. (2006) A survey of motif discovery methods in an integrated framework. *Biol Direct* **1**: 11.

Schneider TD, Stephens RM. (1990) Sequence logos: A new way to display consensus sequences. *Nucleic Acids Res* **18**: 6097–6100.

Sharon E, Lubliner S, Segal E, Stormo G. (2008) A feature-based approach to modeling protein–DNA interactions. *PLoS Comput Biol* **4**: e1000154.

Siddharthan R. (2008) PhyloGibbs-MP: Module prediction and discriminative motif-finding by Gibbs sampling. *PLoS Comput Biol* **4**: e1000156.

Stormo GD, Schneider TD, Gold L, Ehrenfeucht A. (1982) Use of the 'Perceptron' algorithm to distinguish translational initiation sites in *E. coli*. *Nucleic Acids Res* **10**: 2997–3011.

Tang M-HE, Krogh A, Winther O. (2008) BayesMD: Flexible biological modeling for motif discovery. *J Comput Biol* **15**: 1347–1363.

Team RDC. (2008) *R: A Language and Environment for Statistical Computing*, R Foundation for Statistical Computing, Vienna, Australia.

Tompa M, Li N, Bailey TL *et al.* (2005) Assessing computational tools for the discovery of transcription factor binding sites. *Nat Biotechnol* **23**: 137–144.

van Dongen S, Abreu-Goodger C, Enright AJ. (2008) Detecting microRNA binding and siRNA off-target effects from expression data. *Nat Methods* **5**: 1023–1025.

Visel A, Prabhakar S, Akiyama JA *et al.* (2008) Ultraconservation identifies a small subset of extremely constrained developmental enhancers. *Nat Genet* **40**: 158–160.

Visel A, Rubin EM, Pennacchio LA. (2009) Genomic views of distant-acting enhancers. *Nature* **461**: 199–205.

Winick J, Abel T, Leonard MW *et al.* (1993) A GATA family transcription factor is expressed along the embryonic dorsoventral axis in Drosophila melanogaster. *Development* **119**: 1055–1065.

Xie X, Lu J, Kulbokas EJ *et al.* (2005) Systematic discovery of regulatory motifs in human promoters and 3' UTRs by comparison of several mammals. *Nature* **434**: 338–345.

Zhang Y, Liu T, Meyer CA *et al.* (2008) Model-based Analysis of ChIP-Seq (MACS). *Genome Biol* **9**: R137.

Zhu Q, Halfon MS. (2009) Complex organizational structure of the genome revealed by genome-wide analysis of single and alternative promoters in Drosophila melanogaster. *BMC Genomics* **10**: 9.

Chapter 2

R'MES: A Tool to Find Motifs
with a Significantly Unexpected
Frequency in Biological Sequences

Sophie Schbath*,† and Mark Hoebeke*,‡

Statistics of motifs have been widely revisited in the last 15 years due to the increasing availability of genomic sequences. The identification of DNA motifs with biological functions is still a huge challenge of genome analysis. Many functional and essential motifs have the particularity to be very frequent all along the chromosome or to be concentrated in some particular regions or to be preferentially co-oriented with the replication direction. It is therefore neccessary to distinguish significant features (e.g. frequency or skew) from what one could expect just by chance. Many approaches aiming at predicting functional motifs are then based on statistical properties of pattern occurrences in random sequences. R'MES has been initially designed to address the following question: "which are the motifs of length ℓ occurring with an exceptional frequency in this DNA sequence?" Now, it can also detect which motifs have a significant skew in a given sequence and deal with amino acid sequences. Soon it will allow researchers to compare the motif exceptionalities in two different sequences. This chapter presents the R'MES software package from a user's point of view and gives many practical examples, including recent DNA motif identifications.

1. Introduction

The number of complete genomes available in public databases is still increasing and their automatic annotation has become very crucial. The

---

*INRA, UR1077 Mathématique, Informatique et Génome, F-78350 Jouy-en-Josas, France
†Sophie.Schbath@jouy.inra.fr
‡Mark.Hoebeke@jouy.inra.fr

identification of DNA motifs with biological functions in particular organisms remains a huge challenge of this annotation process. Many known functional and essential motifs have statistical properties related to their occurrences along genomes. Some of them are significantly more frequent than what we could expect by chance either all along the chromosomes or in some particular regions of the chromosomes (in front of genes for instance). Others are significantly co-oriented with the replication direction leading to a significant high skew. The prediction of functional motifs is then mostly based on the distinction between expected and unexpected features. R'MES has been initially developped to detect which motifs of a given length occur with an exceptional frequency in a given DNA sequence. R'MES relies on theoretical results on the statistical properties of pattern occurrences in random sequences. The general principle is presented below and the statements of the statistical results are presented in the Methods section. The reader interested by more methodological details and proofs should refer for instance to the book by Robin *et al.* (2005) or to Chapter 6 from Lothaire (2005).

A model as reference. The key idea is to compare the observed count of the motif with the expected one given some knowledge about the sequence. To decide if a word count is indeed unexpected, we need to know what to expect at random. This will be defined by a probabilistic model, i.e. by the description of what "random" means. In practice, Markovian models are used because a Markov chain model of order m fits the observed counts of all oligonucleotides of length 1 up to $(m + 1)$ of the observed sequence. Let us denote by Mm such model.

Choice of the model. Choosing model Mm means to take the base, the dinucleotide, the trinucleotide,..., the $(m + 1)$-mer compositions of the sequence into account to determine what to expect. However, the sequence should be long enough to correctly estimate the 3×4^m parameters of the model (the transition probabilities). Note that a motif of size ℓ can only be analyzed in M0 up to M$(\ell - 2)$ because higher models would fit the motif count itself (the motif will then be expected by definition); Model M$(\ell - 2)$ will be referred to as the maximal model. Since the model determines the reference, changing the reference may change the exceptionality feature of a motif. A word can be exceptionally frequent in one model but expected in another one which, for instance, takes more information on the sequence composition into account. Therefore, when claiming that an observation is statistically significant, it is necessary to mention the reference, i.e. the chosen model.

p-value. To evaluate the significance of the difference between observed and expected counts, we need to evaluate the p-value which is the probability, under our model, to observe as much (or as few) occurrences of our motif of interest. It requires to know the statistical distribution of the count of a motif in Markovian sequences. Several methods exist either to calculate this p-value exactly (not tractable for long sequences) or to approximate it (see Lothaire (2005) or Robin et al. (2005) for some reviews). R'MES implements the two most classical approximations for word count distribution, namely a Gaussian approximation for expectedly frequent words (Prum et al., 1995; Reinert et al., 2000) and a compound Poisson approximation for expectedly rare words (Schbath, 1995; Roquain and Schbath, 2007). The parameters of these approximate distributions are given in the Methods section.

Score of exceptionality. R'MES converts the p-values into scores of exceptionality using the standard one-to-one probit transformation: for a given probability $p \in [0, 1]$, the associated score $u \in \mathbb{R}$ is such that $\mathbb{P}(\mathcal{N}(0, 1) \geq u) = p$. Therefore, exceptionally frequent motifs have high positive scores, whereas exceptionally rare motifs have very negative scores. The Gaussian approximation has the advantage to allow the direct calculation of the scores without going through the p-value calculation, which is much faster. This is not the case for the compound Poisson approximation. Nevertheless, for word counts, the limiting compound Poisson distribution is a Geometric Poisson distribution (a particular case) and efficient algorithms exist to directly compute the tail of such distribution (Nuel, 2008). In the word family case, one needs to compute and to sum up point probabilities from general compound Poisson distributions (Barbour et al., 1992b) which may lead to numerical problems for expectedly frequent families.

Extensions. R'MES can also provide scores associated with the statistical significance of word skews in order to detect significant DNA strand bias. Except for the skew scores, R'MES can analyze frequency motifs in sequences on any alphabets. Markovian models with periodic transition probabilities can also be addressed, which is particularly of interest for coding DNA sequences.

Implementation and availability. R'MES is a free software package available under the GNU General Public License. It can be downloaded with an online user guide from its home web page http://migale.jouy.inra.fr/outils/mig/rmes or directly from http://mulcyber.toulouse.inra.fr/projects/rmes. R'MES comes as a source distribution. It is written in C and C++. Our distribution was specifically designed to be compiled with the GNU GCC compiler. It has been tested on a variety of Unix platforms (Linux, Solaris, MacOS X). R'MES has a companion tool, R'MESPlot, which provides a graphical user interface for the visualization of R'MES generated results.

2. User Guide

R'MES has to be run via a command line which looks like:

```
$rmes [options] -s <filename> -o <string>
```

All the options can be obtained by typing:

```
$rmes --help
```

In this section, we start by giving the most basic use case of R'MES (getting the scores of exceptionality for word frequencies) and then we describe other possible cases with the associated options.

2.1. Getting exceptional frequency scores for words

The most basic use case of R'MES consists in analyzing all the oligonucleotides of a given length in a given sequence. Naturally, the input parameters are:

- *the sequence file*: it is provided after the -s <filename> option; the sequence should be in FASTA format;
- *the word length*: it is provided after the -l <int> option;
- *the order of the model*: it is provided after the -m <int> option;
- *the approximation method*: it is provided either by the --gauss option for the Gaussian approximation or by the --compoundpoisson option for the compound Poisson approximation.

An additional option -o <string> is required to specify the prefix of the output files.

Table 1. Exceptional 6-letter words in *E. coli* complete genome under model M4.

Word	Count	Expect	Sigma2	Score	Rank
ggcgcc	96	2058.166	1181.549	−57.0834	1
gccggc	294	1771.336	943.784	−48.0887	2
ctgcag	958	1982.772	925.338	−33.6882	3
agcgct	779	1773.006	1065.129	−30.4571	4
cggccg	285	858.880	468.397	−26.5164	5
tccgga	906	1708.700	922.467	−26.4288	6
ccgcgg	659	1405.005	847.795	−25.6210	7
gcatgc	589	1145.942	603.744	−22.6664	8
gtcgac	544	1064.316	579.037	−21.6229	9
cagctg	1777	2378.358	793.028	−21.3545	10
...					
agcgcc	2846	1945.233	1136.143	26.7237	4093
ggcgct	2782	1875.940	1107.697	27.2237	4094
gccgga	2622	1744.846	934.807	28.6890	4095
tccggc	2634	1734.641	931.326	29.4701	4096

Example 1. We want to find the exceptional words of length $\ell = 6$ in the complete genome of the bacterium *Escherichia coli* (sequence file ecoli.fasta) under the Markov model of order $m = 4$ (i.e. with respect to the sequence composition in words of length 1 up to 5). For this we first run the following command:

```
$rmes --gauss -s ecoli.fasta -l 6 -m 4 -o ecoli-6.4
```

which will generate the output file named ecoli-6.4.0. The results stored in this file (see below) can be either formatted in a sorted table (see Table 1 for a truncated table) thanks to the rmes.format program included in the distribution (see Sec. 2.5) or displayed via the graphical interface (see Sec. 2.6).

In the sorted table (see Table 1), words are ordered from the most exceptionally rare (score very negative) to the most exceptionally frequent (highest positive score). In this example, one can notice that the most avoided 6-letter words in *E. coli* are palindromic, probably restriction sites (Karlin *et al.*, 1992). The columns named expect and sigma2 refer to the estimation of the expected count and to a variance-like quantity (see Methods section, Eqs. (2) and (4)).

Example 2. We want to find the exceptional words of length $\ell = 9$ in the complete genome of the bacterium *Haemophilus influenzae* (sequence file hinf.fasta) under the Markov model of order $m = 5$ (i.e. with respect to the sequence composition in words of length 1 up to 6). Since 9-letter words are a priori rare in a sequence of length $n \simeq 1.6 \times 10^6$, we use a compound Poisson approximation for the counts. We then run the following command:

```
$rmes --compoundpoisson -s hinf.fasta -l 9 -m 5 -o hinf-9.5
```

and the rmes.format program will produce a sorted list of words whose truncated version is given in Table 2. Note that the two most exceptionally frequent words aagtgcggt and accgcactt are precisely the *uptake* sequences of the bacterium (Smith *et al.*, 1995). Since their expected counts are rather large (greater than 200), it could be more relevant to use a Gaussian approximation for getting their scores. Columns named expect_p and A refer to the estimation of the expected count of clumps and to the overlapping probability (see Methods section, Eq. (7)).

Output file. The above rmes commands produce a unique output file with the ".0" suffix. This file is not intended to be read by the user but only to store the numerical values of each of the quantities of interest (observed counts, expected counts, scores, etc.). To obtain user readable

Table 2. Exceptional 9-letter words in *H. influenzae* complete genome under model M5.

Word	Count	Expect	Expect_p	A	Score	Rank
ttttttttt	8	89.824	62.888	0.29987	−8.8940	1
ttagtgcgg	3	41.810	41.810	0.00000	−7.6622	2
atttttttt	37	104.439	104.439	0.00000	−7.5343	3
aaagtgcga	19	74.081	74.081	0.00000	−7.5331	4
...						
aaaagtgcg	355	136.559	136.559	0.00000	15.5191	262138
agtgcggtc	214	48.116	48.116	0.00000	17.4957	262139
gaccgcact	241	59.489	59.489	0.00000	17.6206	262140
ccgcacttt	614	193.117	193.117	0.00000	24.0413	262141
aaagtgcgg	643	195.573	195.573	0.00000	25.2005	262142
accgcactt	731	220.284	220.284	0.00000	27.0469	262143
aagtgcggt	740	219.081	219.081	0.00000	27.5486	262144

output, this file needs to be formatted. Two tools are available for this: the rmes.format program included in the distribution (see Sec. 2.5) or the Java interface R'MESPlot (see Sec. 2.6). Moreover, the output file will be compressed if the -z option is specified.

Simultaneously analyzing several word lengths. It is possible to simultaneously analyze several word lengths with a single command. For this, the -l <int> option should be replaced with options --lmin <int> and --lmax <int> which specify the minimal and maximal word lengths.

Using the maximal model. When using the maximal model, i.e. when $m = \ell - 2$, the -m <int> option can be replaced with --max; it allows in particular the simultaneous analysis of several word lengths, each one in the associated maximal model (see previous paragraph). Despite its statistical interpretation (see Discussion section), the maximal model allows us to use a more efficient (and thus significantly faster) algorithm to calculate the score from the Gaussian approximation.

Analyzing concatenated sequences. R'MES can consider a concatenation of several sequences as a single sequence. In order to avoid introducing non-existing words at the boundaries of each piece of the concatenated sequence, the latter must be separated by a specific character which depends on the sequence type or alphabet. For DNA, this separator is the letter "Z", and for amino acids the separator is the letter "X". Separators can be either upper or lower case. For user-specified alphabets, the separator must be part of the alphabet definition (see next paragraph).

Warning: the sequence file should however look like a unique sequence file in FASTA format, i.e. with a unique title line and a unique sequence of concatenated bases.

Using different alphabets. The default alphabets included in R'MES are case-insensitive. By default, the nucleotide alphabet is used assuming that the analyzed sequence is a DNA sequence. When analyzing protein sequences, the amino acid alphabet should be used with the --aa option. The user can however specify his/her own alphabet with the --alphabet <character string> option. The character string explicitly describes the alphabet used in the sequence according to a particular format:

validchars:interruptchars:jokerchar

where

- `valdichars` is the list of allowed characters of the sequence,
- `interruptchars` is the list of characters used to separate pieces in a concatenated sequence,
- `jokerchar` is a single character standing for any of the valid characters.

For instance, if a sequence of protein secondary structures is to be analyzed, the alphabet definition could be ABC:X:N, where A could stand for alpha-helix, B for beta-sheet, C for coil and X would be used to separate pieces of sequences and N could replace any of A, B or C.

Note: all characters contained in the sequence but different from the "valid characters" have the same effect than the "separator" character, namely they are not taken into account in the word counting process.

2.2. Getting exceptional frequency scores for word families

To analyze families of oligonucleotides, for instance degenerated oligonucleotides of length 8 with an "n" in the second position, or starting with a purine, or oligonucleotides with their reverse complement,..., the -l <int> option must be replaced by the -f <filename> option, in which <filename> represents the file which enumerates the families (see the format below). Both approximations, Gaussian and compound Poisson, can be used.

Compatibility with other options. The -f <filename> option can be used with the --max (maximal model) option but is not compatible with the word length options -l <int>, --lmin <int> and --lmax <int>.

Format of the family file. A word family has to be composed of words of the same length, say ℓ. All the families contained in a family file have to be composed of the same number of words, say d. The structure of the associated family file is as follows:

- the first line is a title (character string) ended by the # character,
- the second line contains the number of families,

```
4 families rny, rnr, ynr et yry of 16 trinucleotides #
4
16
3
rny
  aac agc acc atc aat agt act att gac ggc gcc gtc gat ggt gct gtt

rnr
  aaa aga aca ata aag agg acg atg gaa gga gca gta gag ggg gcg gtg

ynr
  caa cga cca cta cag cgg ccg ctg taa tga tca tta tag tgg tcg ttg

yny
  cac cgc ccc ctc cat cgt cct ctt tac tgc tcc ttc tat tgt tct ttt
```

Fig. 1. Family file associated to the four families rny, rnr, yny and ynr.

- the third line contains the number d of words in each family,
- the fourth line contains the length ℓ of the words,
- then, each family is listed as follows: the family name followed by all the d words of this family.

As an example, Fig. 1 gives the family file corresponding to the four families rny, rnr, yny and ynr, where r stands for a purine (a or g) and y stands for a pyrimidine (c or t). Note that some family files can be automatically generated thanks to the rmes.gfam program included in the distribution (see Sec. 2.5).

Output file. The rmes command with the -f <filename> option produces an output file (with the ".0" suffix) which contains the list of the word families with their observed count, estimated expected count and score. This list is ordered like in the family file and not with respect to the scores. This output file can also be formatted thanks to rmes.format or visualized with the graphical interface R'MESPlot. The output file will be compressed if the -z option is specified.

Example 3. We want to find the exceptional degenerated words of length $\ell = 8$ with any letter in the second position (i.e. n) in the complete genome (leading strands) of the bacterium *H. influenzae* under the Markov model of order $m = 1$ (i.e. with respect to the sequence composition in bases and dinucleotides). Regarding Example 2, we analyze the sequence file hinf-uptake.fasta in which all occurrences of the *uptake* sequences have been masked (otherwise all subwords of the *uptake* sequences will be found exceptionally frequent). For this, we first generate the file fam.xnxxxxxx which contains the 4^7 word families (from anaaaaaa to tntttttt) and their content by running the rmes.gfam program (see Sec. 2.5). Then we run the following command:

```
$rmes --gauss -s hinf-uptake.fasta -f fam.xnxxxxxx -m 1
      -o hinf-xnxxxxxx.1
```

which will generate the output file named hinf-xnxxxxxx.1.0. As for words, the results stored in this file (observed and expected counts, scores, etc.) can be either formatted in a sorted table (see Table 3 for a truncated table) via the rmes.format program included in the distribution (see Sec. 2.5) or displayed via the graphical interface (see Sec. 2.6).

In this example, one can notice (cf. Table 3) that the most exceptionally frequent family is gntggtgg which is precisely the Chi sequence of *H. influenzae* (Sourice *et al.*, 1998).

Table 3. Exceptional 8-letter degenerated words with an n in the second position in *H. influenzae* complete genome under model M1.

Family	Count	Expect	Sigma2	Score	Rank
tntttttt	845	1642.439	2766.196	−15.1620	1
anatgcat	62	312.138	308.106	−14.2505	2
anttcgaa	32	241.699	239.799	−13.5417	3
antgcatg	29	229.790	227.111	−13.3237	4
tncatgca	32	217.582	217.686	−12.5782	5
tntgcatg	47	240.495	237.758	−12.5488	6
...					
cngaagaa	326	114.194	113.528	19.8787	16379
gnagaaga	270	83.638	85.924	20.1048	16380
tnatcgcc	279	84.555	84.252	21.1840	16381
anatcgcc	288	87.883	87.559	21.3862	16382
anttcatc	469	180.373	178.635	21.5950	16383
gntggtgg	223	55.317	56.352	22.3375	16384

2.3. Analyzing coding DNA sequences

When analyzing coding DNA sequences, it is usually more relevant to use a Markov model which takes the phase into account, i.e. which distinguishes the position of the bases with respect to the codons. More generally, one may be interested to take some periodicity of the sequence into account. For this purpose, the --phases <int> option can be used to specify the period. Note that only the Gaussian approximation is available with phased models. The basic command becomes:

```
$rmes --gauss -s <filename> -l <int> -m <int> --phases
         <int> -o <string>
```

Compatibility with other options. The --phases <int> option can be used with the --max (maximal model) option, with the word length options --lmin <int> and --lmax <int> and with the family option -f <filename>. However, it cannot be used with the --compoundpoisson (compound Poisson approximation) option.

Output files. If a phased model with r phases is used, the above command will produce $r + 1$ output files with suffixes ".1", ".2", ..., ".$(r+1)$". The file suffixed by ".i", $1 \leq i \leq r$, corresponds to the statistical analysis of the number of occurrences ending exclusively on phase i (see Methods section), while the file suffixed by ".$(r+1)$" corresponds to the number of all occurrences (like for a non-phased model). Each of these files has the same structure as the one obtained with a non-phased model (see previous paragraphs). They can be formatted by rmes.format, displayed in the R'MESPlot graphical interface and automatically compressed.

Example 4. We want to find the exceptional words of length $\ell = 8$ in the genes of the bacterium *Escherichia coli* (sequence file ecoli-genes.fasta containing genes separated by the symbols ZZZ to preserve the reading frame) under the three-phased Markov model of order $m = 5$ (i.e. with respect to the sequence composition in bases up to hexamers on each of the three phases). For this we run the following command:

```
$rmes --gauss -s ecoli-genes.fasta -l 8 -m 5 --phases 3
         -o ecoli-genes.8.5
```

Table 4. Exceptional 8-letter words in *E. coli* genes under model M5 with 3 phases.

Word	Count	Expect	Sigma2	Score	Rank
ttctgaca	70	123.926	52.050	−7.4746	1
aaaaaaaa	80	172.531	184.597	−6.8104	2
ttgagctg	138	198.275	115.982	−5.5968	3
aaaaaaac	112	174.200	125.756	−5.5466	4
gtggcggg	101	165.133	137.979	−5.4598	5
...					
aggcgctg	560	447.136	245.921	7.1971	65532
aaaaaata	206	128.751	105.366	7.5256	65533
gattctgg	292	199.572	134.006	7.9844	65534
gctggtgg	783	619.041	391.529	8.2862	65535
ccaccagc	205	117.297	89.722	9.2590	65536

which will generate four output files, one for the number of occurrences ending on each of the three phases and one for the total number of occurrences (suffix ".4"). After formatting the file ecoli-genes.8.5.4, we notice that the two most exceptionally frequent 8-letter words in *E. coli* genes are the Chi sequence of the bacterium (gctggtgg) and its reverse complement (see Table 4).

If we now look separately at occurrences on each phase (by formatting the files ecoli-genes.8.5.1, ecoli-genes.8.5.2 and ecoli-genes.8.5.3), we can notice for instance that most of the 783 occurrences of the Chi sequence occur on phase 1, i.e. like g|ctg|gtg|g with respect to the reading frame, and in a significant way (see Table 5).

2.4. *Getting exceptional skew scores*

When analyzing a DNA sequence, one can be interested to know if an oligonucleotide has a significant skew. The skew of an oligonucleotide is usually defined like the ratio between the oligonucleotide count and the count of its reverse complement (or *conjugate*). The skew is used to describe the strand bias. The *p*-value associated with an oligonucleotide skew can be easily approximated using a Gaussian approximation of the word counts (see Methods section). With R'MES it is then possible to get an exceptionality score of the skew of an oligonucleotide or of a word

Table 5. Exceptional 8-letter words on phase 1 in *E. coli* genes under model M5 with 3 phases.

Word	Count	Expect	Sigma2	Score	Rank
taaactgg	131	192.543	119.364	−5.6330	1
gttctgac	0	19.936	12.851	−5.5610	2
tttctgac	0	17.344	11.870	−5.0342	3
ctggcaaa	111	157.661	89.711	−4.9264	4
. . .					
ctaccaag	12	1.964	1.810	7.4604	65532
gctggtgg	677	532.312	323.696	8.0420	65533
gaggcccc	5	0.326	0.307	8.4290	65534
ccccaagg	10	1.048	0.990	8.9970	65535
gattctgg	257	163.002	104.913	9.1771	65536

family. To do this, the --gauss option from the basic command must be replaced with the --skew option:

```
$rmes --skew -s <filename> -l <int> -m <int> -o <string>
```

Compatibility with other options. The --skew option can be used with the --max (maximal model) option, with the word length options --lmin <int> and --lmax <int> and with the family option -f <filename>. However, it cannot be used with the options --gauss (Gaussian approximation) --compoundpoisson (compound Poisson approximation) and --phases <int> (phased models), neither with the alphabet options --aa and --alphabet <character string>.

Additional output file. The quantities associated with the skew and its significance are stored in an additional file with the ".skew" suffix. This file is presented like a table: the columns successively correspond to the word (or word family), the observed count, the observed count of the reverse complement, the observed skew and the score of exceptionality of the skew.

Example 5. We want to detect the 8-letter words which are the most significantly skewed along both leading strands of the bacterium *E. coli* (ecoli-rep.fasta sequence file). Since we only want to take the base composition of the sequence into account (i.e. $m = 0$), we then run the

Table 6. The 8-letter words which are the most significantly skewed along the leading strands of *E. coli* with respect to the base composition (model M0).

Word	Count	Conj.Count	Skew	Score
tcctccta	13	1	13	8.359953
gggggccc	14	1	14	7.975611
gagtaggg	43	1	43	7.645669
gggtctcc	9	1	9	7.609322
ggggaggg	41	1	41	7.586326

following command:

```
$rmes --skew -s ecoli-rep.fasta -l 8 -m 0 -o ecoli-rep.8.0
```

Table 6 gives an excerpt of the produced file `ecoli-rep.8.0.0.skew`; more precisely it gives the words with the highest scores except those whose conjugate does not occur in the sequence (infinite skew in this case). One can recognize that one of them (ggggaggg) is a member of the KOPS family (Bigot *et al.*, 2005).

2.5. *Utilities*

`rmes.format` This program displays the results contained in an output file (`<rmesfilename>`) generated by the `rmes` command. It produces a table with the motifs sorted according to their exceptionality scores. The basic command is:

```
$rmes.format < <rmesfilename> > <tablefilename>
```

Table 1, for instance, is an excerpt of the table produced by the command:

```
$rmes.format < ecoli-6.4.0 > ecoli-6.4.0.table
```

The meaning of the different columns of the output table depends on the approximation used, i.e. from the options `--gauss`, `--compoundpoisson`, or `--skew`.

- Gaussian approximation (options `--gauss` or `--skew`): the six columns successively correspond to the motifs, their observed count, their estimated expected count, their estimated limiting variance, their score of exceptionality and their rank when all motif scores are sorted by increasing order.

- Compound Poisson approximation (option `--compoundpoisson`): the seven columns successively correspond to the motifs, their observed count, their estimated expected count, the estimation of their expected number of clumps (see Sec. 5.5), their overlapping probability, their score of exceptionality and their rank when all motif scores are sorted by increasing order.

If the input file contains results for more than one word length, then as many tables as the considered word lengths will be produced. However, if only a subset of word lengths (between ℓmin and ℓmax) is of interest, the `-i <int>` and `-a <int>` options can be used (`-l <int>` is enough for a unique word length).

The significance associated with a given score is obtained thanks to the standard Gaussian distribution with mean 0 and variance 1. In particular, scores ranging from -3 to 3 are not really significant (p-value greater than 0.00135). Therefore, motifs with such scores are removed by default from the tables. To specify other thresholds, one can use the `--tmax <float value>` and `--tmin <float value>` options; in this case, only motifs with a score greater than the maximal threshold or less than the minimal threshold will be displayed in the table.

`rmes.gfam` This program allows us to generate family files when the corresponding families are degenerated DNA motifs which can be written thanks to the bases a, c, g, t and n. The basic command is:

```
$rmes.gfam  -t <label> -p <string>
```

The `-t <label>` option just specifies the title of the resulting family file (this title will be the first line of the family file).

The pattern specified by the `-p <string>` option is the template to generate the degenerated motifs (families). Its length ℓ will be the length of the words in the families. Each of its ℓ characters can take a value among #, a, c, g, t and n.

- If the i-th character of the template is a (respectively c, g or t), each family is labeled with a pattern whose i-th character is a (resp. c, g or t) and all words from all families will have an a (resp. c, g or t) at position i.

- If the i-th character is n, each family is labeled with a pattern whose i-th character is n and a series of words belonging to a same family will be generated whose i-th character is a letter of the alphabet.
- If the i-th character is #, a series of families will be generated, each family being labeled with a pattern whose i-th character is a letter of the alphabet.

The number of families is then 4^α where α is the number of #'s in the template and the number of words in each family is 4^β where β is the number of n's in the template.

For instance, the family file fam.xnxxxxxx from Sec. 2.2 which contains the 4^7 degenerated octamers having an n in the second position has been produced by the following command:

```
$rmes.gfam -t 'Octamers of the form xnxxxxxx' -p '#n######'
          >fam.xnxxxxxx
```

2.6. *Graphical user interface*

As already stated above, the raw output files generated by R'MES are not destined to be read directly. The previous section describes a command-line utility for reformatting these result files, but there also exists a graphical user interface that can be used to explore the contents of R'MES's output, called R'MESPlot .

Exploring data. At a basic level, R'MESPlot is capable of displaying the contents of any number of result files as a tree. Each top-level branch contains a subtree with results relative to the same sequence. The subtree itself is subdivided according to the approximation method, the phase, the word length and the Markov order, as can be seen in Fig. 2. Selecting tree leaf node will add a table with its contents to the data panel (this table is similar to the one produced by the rmes.format utility). Each table row represents the results for a single word, and the columns are its attributes. The latter can be independent of the approximation (like the rank, the word itself, the observed count, the score) or specific (the estimated variance for Gaussian approximations for example). As can be expected, the whole

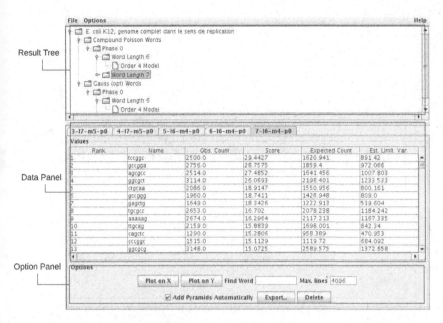

Fig. 2. Screenshot of R'MESPlot showing the result tree and the data table.

table can be sorted according to any of the column headers. Moreover, an option panel allows the user to look for a specific word or to define which graphical representation to build, either from the whole result set or from a given word.

Drawing comparative plots. Sometimes it may be desirable to compare, at a glance, word scores from two different result sets. In order to do so, all that is needed is to select them from the result tree and to use the option panel to define on which axis each of the dataset will be displayed. The resulting plot will contain tiny cross symbols, one for each word. This may for instance occur when a word has an infinite or "not a number"-like score in one of the sets. The application uses shades of green Figure 3 shows a plot built from two result sets. Selecting a data point in the plot by clicking on a cross will switch the display of the corresponding word on or off. Other graphical manipulations are allowed with the option panel below the drawing area.

Building pyramids. In some cases it might be relevant to examine if the exceptionality of a word might be caused by the exceptionality of some of its subwords. R'MESPlot allows this, if result data are available

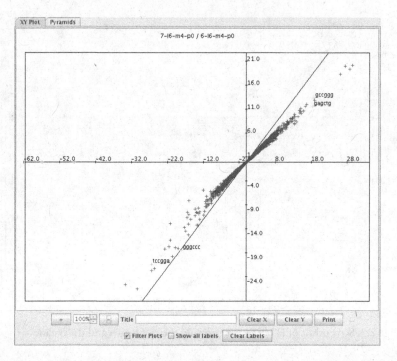

Fig. 3. Screenshot of R'MESPlot showing how to plot word scores from two different result sets. Along the x-axis: scores computed using the Gaussian approximation. Along the y-axis: scores computed using the compound Poisson approximation. Some of the words have been highlighted with a mouse click.

for different word lengths related to a same sequence in a single result file. Once the file has been loaded into R'MESPlot, one has to select a specific word length and, in the data table, a specific word. The "Pyramids" tab of the graphics window will show the exceptionality scores of the selected word and of all its subwords: the top square corresponds to the actual word, the squares of the layer below correspond to the scores of the two subwords whose length is one less than the selected word, and so forth. Square colors are indicative of the score level of each of the words. The application uses shades of green (resp. red) to highlight words with exceptionally high (resp. low) scores. Figure 4 shows a representation of word pyramids. To enhance the readability of this printed greyscale figure, a "+" (resp. "−") symbol was drawn on squares with exceptionally high (resp. low) scores. For example, word tccggc has a high exceptionality score (vivid green top square of left pyramid), as is the case for its left

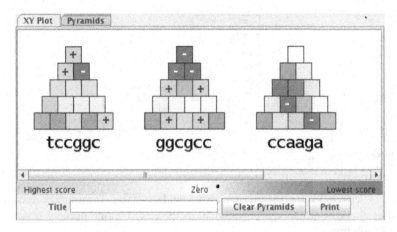

Fig. 4. Screenshot of R'MESPlot displaying pyramids of words to assess their exceptionality with respect to the exceptionality of their subwords.

subword of length 5 (tccgg, leftmost square of the layer just below the topmost layer). Its right subword of length 5 (ccggc, rightmost square of the same layer) however has an exceptionally low score, as denoted by its deep red color.

2.7. Implementation details

This section starts by detailing the design of R'MES's main classes and its core algorithms for one of the available approximations, followed by some considerations about the time and space complexities with respect to the word and sequence lengths as well as the Markov order of the model. Finally, a few tables summarize measurements of both execution time and memory footprint for a set of runs of R'MES.

2.7.1. Main data structures and algorithms

Design of main classes in R'MES. Basically, R'MES computes various quantities on words present in a sequence. A significant part of the program thus relies on the extraction of subwords from larger words or from the whole sequence, or the creation of new words by adding letters to already existing ones. In order to be capable of uniformly using word and sequence objects to perform these operations, they are provided by an

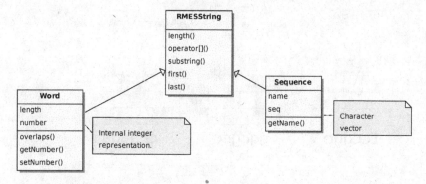

Fig. 5. Hierarchy of classes representing words and sequences. The use of an abstract base class allows the definition of polymorphic methods to uniformly manipulate single letters or substrings.

abstract RMESString class from which both the Word and the Sequence classes are derived. Indeed, in the current release of R'MES, the internal representation of words (coded as long integers) and sequences (coded as character vectors) are different enough to prevent the use of a single class for both entities. Figure 5 shows how these classes are organized.

It also comes as no surprise that the main data structure used to store the results (the ResultSet class) is made of a series of vectors, one for each word-related quantity (count, expected count, variance, score and so on), indexed by an integer representation of each word. The exact nature of the quantities varies with the type of approximation (Gaussian or compound Poisson), leading to a C++ class hierarchy reflecting this diversity, as shown in Fig. 6. The duality of the hierarchy is needed to distinguish results related to word families from those dealing with raw words.

A similar, although not identical, structural approach has been taken for the organization of estimation related classes. Each combination of approximation type and word or family results gives rise to a specific class, as shown in Fig. 7. The only method present in the root of the estimator classes is the actual estimation method, which is over-written by each sub-class with its specific algorithm. It is the task of a dedicated factory class to instantiate the correct estimator given the combination of command line arguments.

Apart from these core class hierarchies, R'MES obviously has also some more trivial classes for sequence reading, word counting and processing

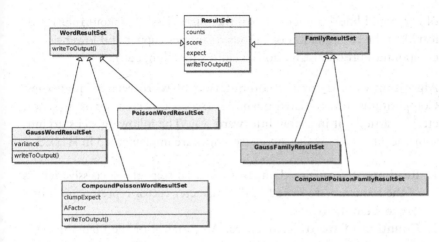

Fig. 6. Hierarchy of classes storing computational results. Examples of class specific attributes and over-written methods are shown for word related result classes. Their counterparts for family-related result classes are omitted.

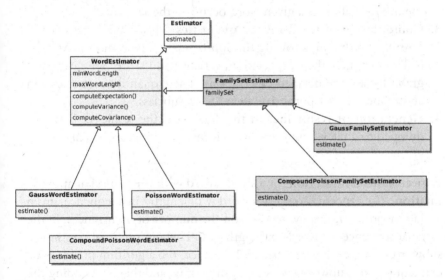

Fig. 7. Hierarchy of estimation performing classes. Examples of class specific attributes and over-written methods are shown for word-related estimation classes and family-related estimation classes. Moreover, computation of variance and covariance for two words is sufficiently approximation-independent to be declared at the WordEstimator level. There is no formal score computation method, because its signature would be too divergent between approximations. Hence, score computation is directly performed in the estimate() method where the needed quantities are available.

of command-line arguments. Also, there is a class (StatRoutines) dedicated to the computation of various quantities (upper and lower tails of compound Poisson distributions, Gamma function, etc.).

Algorithm outline for the computation of word-related quantities. Computing word-associated quantities (score, expected count, variance, etc.) is carried out in a straightforward way. The following steps go into some detail as to how the successive stages are implemented in R'MES.

1. **Reading of sequence data.** In this trivial step, the successive letters of the sequence are stored in the character vector representation of a Sequence instance.
2. **Counting of word occurrences.** An instance of the Counter class, created by specifying the lengths of the words to be counted, scans the Sequence class, and tallies the subsequences of the correct length at each position. As a result, the Counter instance is capable of returning the number of times a given word occurs in the sequence.
3. **Computation of word-related quantities.** This step iterates over all possible words and, knowing the approximation type and the Markov order, computes their expectation, variance and score (see next paragraph for a more detailed description). These quantities are stored in an instance of the appropriate ResultSet subclass.
4. **Generation of result file.** In this final step, the result set is simply dumped to a file whose prefix was given as an initial argument.

Expectation computation for a single word with the Gaussian approximation. The expected count of a word $\mathbf{w}$ of length ℓ in a sequence, given a Markov order m, mainly depends on the number of occurrences, in the original sequence, of words of length $m + 1$ whose prefix of size m is a subword of $\mathbf{w}$ (see Eq. (2), Sec. 5.2). Hence, the algorithm proceeds by extracting every subword ($\mathbf{w}'$) of length m of $\mathbf{w}$, and then by building the set of all possible words by adding one of the letters from the alphabet to the right of $\mathbf{w}'$. A step by step pseudo-code of the algorithm can be found in Listing 1, and is an implementation of Eq. (2). The algorithm for computing the variance follows a similar approach although it is a bit more complex due to the fact that self-overlaps of $\mathbf{w}$ have to be taken into account and counts of subwords have to be done also inside the word itself (see Eq. (4)).

Listing 1: Pseudo-code for expectation computation

```
//Extract lefmost word of length m+1 from word w.
prefix=w.substring(0,m)
//Initialize expectation with number of occurrences of prefix
//in whole sequence. 'c' is an instance of Counter containing
//all word counts for the current sequence.
expectation=c.wordCount(prefix)
//Loop on all possible subwords of length m+1 in w
for (p=1; p<w.length()-m; ++p) {
    //Start by extracting subword of length m at position p
    runningprefix=w.substring(p,p+m-1)
    //Keep track of occurrences of all possible subwords
    quantity=0
    //Loop on all letters of the alphabet
    for (i=0;i<alphabet.size();++i) {
        //Build word of length m+1 by appending letter to subword of length m
        subword=runningprefix+alphabet[i]
        //Add occurrences of subword in the sequence
        quantity+=c.wordCount(subword)
    }
    //Finally get occurrences of actual subword of length m+1 at position p in w.
    actualsubword=runningprefix+w[p+m]
    //And compute the expectation.
    expectation*=c.wordCount(actualsubword)/quantity
}
```

2.7.2. Space and time complexity

In the current implementation of R'MES, both time and space complexity are determined first and foremost by the word length (ℓ) and the size of the alphabet (k). The size of the alphabet determines the number of bits b needed to store a single character, with $b = log_2(k)$. The amount of memory needed to store all possible integer representations of words of length ℓ then equals $2^{b\ell}$. If q is the number of quantities associated with each word, and s the space taken by a single quantity, the overall space complexity becomes $O(qs \times 2^{b\ell})$ which explains the present limitation[a] of $\ell \leq 14$.

The time complexity also depends on the same set of parameters with the added cost of the approximation type (Gaussian or compound

---

[a]We are in the process of changing the internal representation of words to overcome this limitation on the word length.

Poisson). So globally, the time complexity is about $O(c \times 2^{b\ell})$, where c is related to the approximation and the Markov order. In some cases, c may be relatively small (e.g. Gaussian approximation + maximal model). At the opposite, compound Poisson approximation for short words may lead to a high value of c (see next paragraph).

In these space and time complexity expressions, the costs related to initially processing the sequence and storing it in memory have been neglected. Note however that when collecting statistics on small words in huge sequences, this assumption may not hold, leading to space and time complexities of $O(n)$ where n is the length of the sequence.

2.7.3. Computation time and memory requirements

Tables 7 and 8 contain execution times (in seconds) for R'MES (respectively for the Gaussian and the compound Poisson approximations) with different word lengths (2 to 10) and Markov orders (0 to the maximal order for the given word length). The sequence file used to measure execution time was the complete genome of *E. Coli* along its direction of replication (≈ 4.7 Mb). The test platform had Ubuntu Linux 9.10 installed on an Intel Pentium Core2 Duo CPU at 2.4 Ghz.

One can see that the Gaussian approximation is very fast for short words ($\ell \leq 8$) with an increasing runtime as the word length (and thus

Table 7.   R'MES runtime (in seconds) using the Gaussian approximation for different word lengths ℓ and different orders m for the Markov model.

	Word length ℓ								
m	2	3	4	5	6	7	8	9	10
0	≤ 1	≤ 1	≤ 1	≤ 1	≤ 1	≤ 1	1.01	2.63	8.59
1		≤ 1	≤ 1	≤ 1	≤ 1	≤ 1	1.08	2.80	8.95
2			≤ 1	≤ 1	≤ 1	≤ 1	1.10	3.06	9.94
3				≤ 1	≤ 1	≤ 1	1.12	2.95	10.27
4					≤ 1	≤ 1	1.17	3.14	10.80
5						≤ 1	1.31	3.75	13.13
6							1.18	7.16	27.28
7								2.20	80.05
8									5.61

Table 8. R'MES runtime (in seconds) using the compound Poisson approximation for different word lengths ℓ and different orders m for the Markov model.

	Word length ℓ								
m	2	3	4	5	6	7	8	9	10
0	36,186	9,095	3,757	1,079	266.25	74.00	16.91	6.80	10.98
1		9,014	3,739	1,071	265.58	74.04	16.86	6.76	10.90
2			734	378	85.78	18.41	6.41	4.01	8.85
3				64.52	34.83	6.91	2.20	3.17	8.48
4					4.49	3.70	1.62	2.77	8.09
5					≤ 1	1.34	2.68	8.21
6							1.11	2.65	8.09
7								2.60	7.96
8									7.80

the number of analyzed words) or the model order increases. For long words, one can observe the benefit of using the maximal model. The picture is different for the compound Poisson approximation which is drastically penalized by short words simply because their observed and expected counts are high (this approximation is theoretically not adapted in these cases). Runtime is quite small for long words and decreases as the model order increases (only word periods less than $\ell - m$ have indeed to be considered, see Methods section).

Tables 9 and 10 show the amount of memory (in Mb) needed by R'MES for the same parameters as those used to measure execution times.

3. Discussion

We will first illustrate how R'MES could be useful to identify functional DNA motifs, then discuss the choice of the Markov model and finally the choice of the approximation for the word count distribution.

DNA motif discovery. We describe here the strategy used in Halpern *et al.* (2007) to identify the Chi motif in the bacteria *Staphylococcus aureus*. This identification is based on the following properties of already known Chi motifs for several bacteria (El Karoui *et al.*, 1999): they are very frequent and their occurrences are mainly oriented in the direction of

Table 9. R'MES memory footprint (in Mb) using the Gaussian approximation for different word lengths ℓ and different orders m for the Markov model.

	Word length ℓ								
m	2	3	4	5	6	7	8	9	10
0	26.0	26.0	26.0	26.1	26.2	26.5	27.9	33.4	55.4
1		26.0	26.0	26.1	26.2	26.5	27.9	33.4	55.4
2			26.0	26.1	26.2	26.5	27.9	33.4	55.4
3				26.1	26.2	26.5	27.9	33.4	55.4
4					26.3	26.5	27.9	33.4	55.4
5						26.8	27.9	33.4	55.5
6							29.0	33.5	55.6
7								37.9	55.9
8									73.5

Table 10. R'MES memory footprint (in Mb) using the compound Poisson approximation for different word lengths ℓ and different orders m for the Markov model.

	Word length ℓ								
m	2	3	4	5	6	7	8	9	10
0	32.2	28.0	26.7	26.3	26.3	26.7	28.4	35.5	63.8
1		28.0	26.7	26.3	26.3	26.7	28.4	35.5	63.8
2			26.6	26.3	26.3	26.7	28.4	35.5	63.8
3				26.3	26.3	26.7	28.4	35.5	63.8
4					26.3	26.7	28.4	35.5	63.8
5						26.7	28.5	35.5	63.8
6							28.5	35.6	63.9
7								35.8	64.1
8									64.9

DNA replication. The first step has been to extract the backbone[b] of the *S. aureus* genome by comparing the genome of six strains of the bacteria.

---

[b]The backbone is composed of the genomic regions common to all the compared strains.

The obtained backbone contains about 2.44×10^6 letters and can be retrieved from the MOSAIC database[c] (Chiapello *et al.*, 2005). The second step was to search for motifs which are frequent enough, exceptionally frequent and relatively skewed on the leading strands. They start by analyzing 8-letter words (like for the Chi motif of *E. coli*) with R'MES but none of the most over-represented and skewed motifs were frequent enough to be retained as potential Chi candidates. They thus focused on 7-letter words. Scores of exceptionality were calculated with the Gaussian approximation and in the maximal model, namely model M5. Six motifs had an exceptionality score greater than 11 (see Table 11 or Fig. 8 for a global view). Two of them have a negative skew score so they were not retained. A biological experiment has then been done to test for *S. aureus* Chi activity of the four candidates: gaaaatg, ggattag, gaagcgg and gaattag. The conclusion was that gaagcgg is necessary and sufficient to confer Chi activity in *S. aureaus*. This strategy has also been successfully used to predict and validate the Chi motif of three species of the *Streptococcus* genus (Halpern *et al.*, 2007).

Table 11. The 10 most exceptionally frequent 7-letter words under model M5 in the *S. aureaus* complete genome. Columns correspond respectively to the word, its observed count, its estimated expected count, its normalizing factor, its score of over-representation under model M5, its observed skew and its skew score under model M0.

w	$N_{obs}(w)$	$\widehat{\mathbb{E}}_5[N(w)]$	$\widehat{\sigma}_5^2(w)$	$u_5(w)$	Skew	Score
taaaaaa	1542	1214.3	603.4	13.34	1.61	−1.28
gaaaatg	1067	789.9	454.2	13.00	2.48	1.13
taaaatt	1356	1062.6	552.8	12.48	1.04	−1.53
ggattag	266	143.2	97.5	12.43	2.53	1.52
gaagcgg	272	162.4	88.1	11.67	7.56	2.91
gaattag	614	420.7	274.4	11.67	3.89	7.23
gaaaaag	1177	942.1	518.0	10.32	3.52	2.53
taagatt	316	201.3	130.9	10.03	1.07	−2.98
ttaaaag	1059	856.5	431.6	9.75	2.00	3.85
gatttag	657	488.1	305.9	9.66	2.16	4.25

---

[c]http://genome.jouy.inra.fr/mosaic/

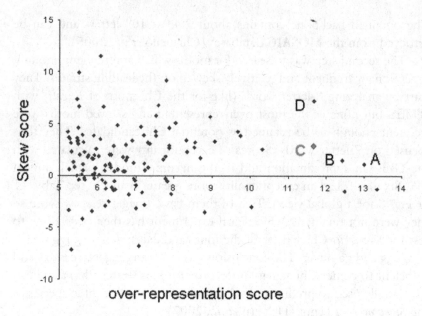

Fig. 8. Over-representation scores under M5 and skew scores under M0 for the most over-represented 7-letter words (over-representation scores greater than 5) in the backbone genome of *S. aureus*. The four best candidates (motifs A to D) are indicated. Motif C (gaagcgg) has been confirmed as the functional Chi site of *S. aureus*.

Choice of the model. Clearly the exceptionality score depends on the chosen Markov model (i.e. on the order m of the model, but also on the phased/unphased feature) because it quantifies how much the observed count of a given motif differs from the count one should expect under the model. Recall that a Markov model of order m fits the sequence composition in short words of length 1 up to $(m + 1)$. Therefore, when the order of the Markov model increases, the model better fits the sequence composition and less exceptional words are globally found. This is illustrated by the boxplots of Fig. 9. Moreover, in a high-order model we have a more accurate knowledge about the sequence composition than in a low-order model: the significance of a word frequency has then no reason to be the same. This point is illustrated by the plot of Fig. 9 which compares scores of 8-letter words in models M1 and M6 in the *E. coli* genome. We recognize the Chi motif gctggtgg which is clearly outside the cloud but let us take the case of the word ggcgctgg. It occurs 761 times in the sequence; it has a significantly high score of 62.4 in model M1 (it is the second most

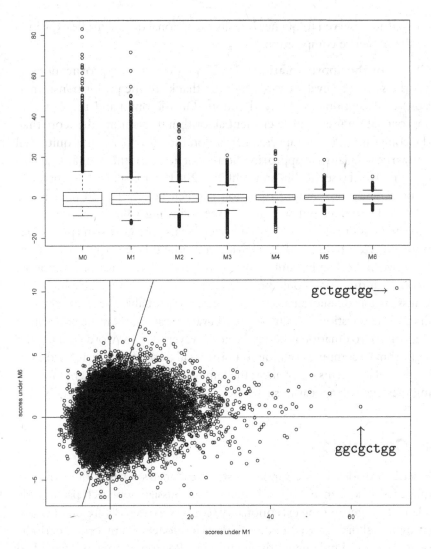

Fig. 9. Exceptionality scores for the 65 536 eight-letter words in the *E. coli* genome. Top: Boxplots of the scores under models M0 to M6. Bottom: Scores under models M1 (*x*-axis) and M6 (*y*-axis).

exceptional word) but has a score of 0.8 in model M6 (rank 17100). It simply means that its high frequency can be explained by the sequence composition of 7-letter words; indeed it is expected about 749 times in model M6. The maximal model ($m = \ell - 2$) is probably the best one to check if an observed frequency is really exceptional or can just be explained by the sequence composition.

Choice of the approximation. R'MES proposes two approximations to get the score (equivalent to the p-value thanks to the probit transformation) of exceptionality of word counts. On the one hand the Gaussian approximation uses a more efficient algorithm to compute the score than the compound Poisson approximation; on the other hand, it is quite well known that a Gaussian approximation is not good for the count of expectedly rare words (see Robin and Schbath (2001) for instance). In this case, a compound Poisson approximation is much better. The frontier is not easy to determine, but we can use the following rule of thumb: if the expected count is greater than 100, then we use the Gaussian approximation; if it is less than 20, then we use a compound Poisson approximation. In between, both approximations mostly give the same list of exceptional words (but the scores could be different). Since the scores obtained by the Gaussian approximation are faster to get, it is probably the method to try first. If the question is really to get accurate p-values, then the compound Poisson approximation is better. Note however that, for word families, the algorithm to compute compound Poisson tail distributions may encounter numerical problems especially if the family is expectedly frequent (a warning message will appear in such cases).

4. Conclusion

R'MES provides a wide variety of statistical analyses based on word frequencies in sequences. It implements the Gaussian approach particularly adapted for getting the exceptionality scores of expectedly frequent words (roughly "short" words) under Markovian models of any order, periodic or not. It also implements the compound Poisson approach devoted to expectedly rare words in homogeneous Markovian models of any order. For both approaches, R'MES can consider single words and families of words.

Several extensions or improvements are currently addressed, some related to R'MES's core functionalities, and others in the user interface field. First of all, R'MES will allow the comparison of exceptionalities between two sequences. In order to do so, the p-values proposed by Robin *et al.* (2007) will be computed. Also, the internal representation of words is undergoing a major overhaul to lift the word limitation barrier. And on a very technical level, the next major release of R'MES will rely on new packaging tools, allowing us to release the software on a wider range of platforms, including the Windows™operating system. In this effort to widen the community of R'MES users, the potential complexity of its command-line leads us to initiate, or to plan, the development of more user friendly interfaces. In a first stage, a graphical user interface will offer an easier way to specify the arguments and to launch the actual R'MES program. Needless to say, this GUI will embed R'MESPlot giving access to execution and result exploration in a single application. As installing a stand-alone application can still be cumbersome, a Web interface is also planned. It will offer remote execution of a trimmed-down version of R'MES (to prevent resource exhaustion on the hosting platform) with a Web browser.

5. Methods

In this section, we give the expression of most of the quantities calculated by R'MES and relate them to theoretical results on word occurrences in Markovian sequences. First, we define the Markov chain models used in R'MES and their estimation from an observed sequence. Next we give the formula for the estimated expected count and we state the Gaussian approximation for the count distribution. In particular, we give the expression of the estimated variance. We also derive the exceptionality score for the skew. Finally, we present the compound Poisson approximation for the count distribution which relies on the fundamental fact that words occur in disjoint clumps of overlapping occurrences, the number of clumps being approximately Poisson distributed.

Let us consider a random sequence $S = X_1 X_2 \cdots X_n$ of length n on a given alphabet $\mathcal{A}$, i.e. $X_i \in \mathcal{A}$ for $i = 1, \ldots, n$. Without loss of generality, let us suppose that $\mathcal{A} := \{a, c, g, t\}$.

Let us define a word $\mathbf{w} = w_1 w_2 \cdots w_\ell$ of length ℓ and $\mathcal{W} = \{\mathbf{w}_1, \mathbf{w}_2, \ldots, \mathbf{w}_r\}$ a word family composed of r words of length ℓ. We are then interested in approximating the p-values $\mathbb{P}(N(\mathbf{w}) \geq N^{\text{obs}}(\mathbf{w}))$ and $\mathbb{P}(N(\mathcal{W}) \geq N^{\text{obs}}(\mathcal{W}))$ where $N(\cdot)$ and $N^{\text{obs}}(\cdot)$ stand respectively for the random number of occurrences in S and the observed number of occurrences in the DNA sequence.

5.1. Markov chain models

The random sequence $\mathbf{S} = X_1 X_2 \cdots X_n$ is a stationary Markov chain of order m if and only if the distribution of X_i conditionally to all the previous letters is equal to the distribution of X_i conditionally to the only m previous letters $X_{i-m}, X_{i-m+1}, \ldots, X_{i-1}$. It simply means that the letters of the sequence are dependent of each other but they only depend on the m previous letters. Such a model, denoted by $\mathbf{M}m$, is thus defined by the following transition probabilities:

$$\pi(a_1 a_2 \cdots a_m, b)$$
$$= \mathbb{P}(X_i = b \mid X_{i-m} = a_1, X_{i-m+1} = a_2, \ldots, X_{i-1} = a_m),$$
$$\forall a_i, \ b \in \mathcal{A}$$

and an initial distribution for the first m letters $X_1 X_2 \cdots X_m$. Classically the initial distribution is set to the stationary distribution μ of the Markov chain so we have

$$\mu(a_1 a_2 \cdots a_m) = \mathbb{P}(X_1 = a_1, X_2 = a_2, \ldots, X_m = a_m), \quad \forall a_i \in \mathcal{A}.$$

Estimation of the parameters. To estimate the transition probabilities given an observed sequence, it is classical to maximize the likelihood of the observed sequence. By doing so, we get the following estimates:

$$\widehat{\pi}(a_1 a_2 \cdots a_m, b) = \frac{N^{\text{obs}}(a_1 a_2 \cdots a_m b)}{N^{\text{obs}}(a_1 a_2 \cdots a_m +)}$$

with $N^{\text{obs}}(a_1 a_2 \cdots a_m +) = \sum_b N^{\text{obs}}(a_1 a_2 \cdots a_m b)$. Moreover, the usual estimate of $\mu(a_1 a_2 \cdots a_m)$ is $\widehat{\mu}(a_1 a_2 \cdots a_m) = N^{\text{obs}}(a_1 a_2 \cdots a_m)/(n - m + 1)$.

It is not difficult to show that in a Markov chain of order m whose parameters are set to the previous estimates, we have:

$$\mathbb{E}N(a_1 a_2 \cdots a_m a_{m+1}) = (n - m)\widehat{\mu}(a_1 a_2 \cdots a_m)\widehat{\pi}(a_1 a_2 \cdots a_m, a_{m+1})$$
$$\simeq N^{\text{obs}}(a_1 a_2 \cdots a_m a_{m+1});$$

in other words, the random sequences will have in average the same composition as the observed sequence in words of length $m + 1$, and by recursion in words of length m down to 1.

Phased Markov chain for coding DNA sequences. In a phased Markov chain of order m, say with three different phases and denoted by Mm_3, we have three different transition probabilities from $a_1 a_2 \cdots a_m$ to b depending on the phase of the letter that will be set to b. Namely, we have

$$\mathbb{P}(X_i = b \mid X_{i-m} = a_1, X_{i-m+1} = a_2, \ldots, X_{i-1} = a_m)$$

$$= \begin{cases} \pi_1(a_1 a_2 \cdots a_m, b) & \text{if } i\%3 = 1, \\ \pi_2(a_1 a_2 \cdots a_m, b) & \text{if } i\%3 = 2, \\ \pi_3(a_1 a_2 \cdots a_m, b) & \text{if } i\%3 = 3. \end{cases}$$

The estimates are

$$\widehat{\pi}_\phi(a_1 a_2 \cdots a_m, b) = \frac{N^{\text{obs}}(a_1 a_2 \cdots a_m b, \phi)}{N^{\text{obs}}(a_1 a_2 \cdots a_m +, \phi)},$$

where $N^{\text{obs}}(\mathbf{w}, \phi)$ denotes the number of occurrences of $\mathbf{w}$ ending into phase ϕ.

5.2. *Estimated expected counts*

The number of occurrences $N(\mathbf{w})$ of an ℓ-letter word $\mathbf{w}$ in the sequence $\mathbf{S} = X_1 X_2 \cdots X_n$ can simply be defined by

$$N(\mathbf{w}) = \sum_{i=1}^{n-\ell+1} \varUpsilon_i(\mathbf{w}), \qquad (1)$$

where $\varUpsilon_i(\mathbf{w})$ equals 1 if and only if an occurrence of $\mathbf{w}$ starts at position i in the sequence and 0 otherwise. This leads to $\mathbb{E}_m[N(\mathbf{w})] = \sum_{i=1}^{n-\ell+1} \mathbb{E}_m[\varUpsilon_i(\mathbf{w})]$ where $\mathbb{E}_m$ denotes the expectation under model Mm. Note that $\mathbb{E}_m[\varUpsilon_i(\mathbf{w})] = \mathbb{P}(\mathbf{w} \text{ occurs at position } i)$ and thanks to the Markov property, we have

$$\mu_m(\mathbf{w}) := \mathbb{P}(\mathbf{w} \text{ occurs at position } i)$$

$$= \mu(w_1 \cdots w_m)\pi(w_1 \cdots w_m, w_{m+1}) \times \cdots \times \pi(w_{\ell-m} \cdots w_{\ell-1}, w_\ell).$$

Therefore, we have $\mathbb{E}_m[N(\mathbf{w})] = (n - \ell + 1)\mu_m(\mathbf{w})$.

If we now replace the model parameters by their estimates, we get the following estimator for the expected count of $\mathbf{w}$:

$$\widehat{\mathbb{E}}_m N(\mathbf{w}) = \frac{\prod_{j=1}^{\ell-m} N(w_j \cdots w_{j+m-1} w_{j+m})}{\prod_{j=2}^{\ell-m} N(w_j \cdots w_{j+m-1} +)}. \tag{2}$$

This is the expect quantity calculated by R'MES.

Generalization to phased models. Under model Mm_3, we have

$$\widehat{\mathbb{E}}_{m\_3} N(\mathbf{w}, \phi) = \frac{\prod_{j=1}^{\ell-m} N(w_j \cdots w_{j+m-1} w_{j+m}, \phi - \ell + j + m)}{\prod_{j=2}^{\ell-m} N(w_j \cdots w_{j+m-1}+, \phi - \ell + j + m)}$$

and $\widehat{\mathbb{E}}_{m\_3} N(\mathbf{w}) = \sum_{\phi=1}^{3} \widehat{\mathbb{E}}_{m\_3} N(\mathbf{w}, \phi)$.

Generalization to word families. We simply have

$$\widehat{\mathbb{E}}_m N(\mathcal{W}) = \sum_{\mathbf{w} \in \mathcal{W}} \widehat{\mathbb{E}}_m N(\mathbf{w}).$$

5.3. *Gaussian approximation*

The count $N(\mathbf{w})$ is a sum of $(n - \ell + 1)$ random Bernoulli variables $\Upsilon_i(\mathbf{w})$ with mean $\mu(\mathbf{w})$. Since these Bernoulli variables are not independent, the count does not follow a binomial distribution. By using a Central Limit Theorem for Markov chains, the asymptotic normality of the count can be established:

$$\frac{N(\mathbf{w}) - \mathbb{E}_m[N(\mathbf{w})]}{\sqrt{\mathbb{V}_m[N(\mathbf{w})]}} \xrightarrow{\mathcal{D}} \mathcal{N}(0, 1) \quad \text{as } n \to +\infty, \tag{3}$$

where $\mathbb{V}_m$ denotes the variance under model Mm. However, if we replace the expected count $\mathbb{E}_m[N(\mathbf{w})]$ by its estimate $\widehat{\mathbb{E}}_m N(\mathbf{w})$, the plug-in estimator of the variance is no more the good normalizing factor to get an asymptotic variance equal to 1.

Estimated variance. The score of exceptionality calculated by R'MES is the following asymptotically Gaussian ratio (Prum *et al.*, 1995; Reinert *et al.*, 2000):

$$u_m(\mathbf{w}) := \frac{N(\mathbf{w}) - \widehat{\mathbb{E}}_m[N(\mathbf{w})]}{\sqrt{\widehat{\sigma}_m^2(\mathbf{w})}} \xrightarrow{\mathcal{D}} \mathcal{N}(0, 1),$$

where

$$\widehat{\sigma}_m^2(\mathbf{w}) = \widehat{\mathbb{E}}_m[N(\mathbf{w})] + 2 \sum_{d=1}^{\ell-m-1} \varepsilon_{\ell-d}(\mathbf{w}) \widehat{\mathbb{E}}_m[N(w_1 \cdots w_d w_1 \cdots w_\ell)]$$

$$+ \{\widehat{\mathbb{E}}_m[N(\mathbf{w})]\}^2 \left[\sum_{a_1,\dots,a_m} \frac{[N_\mathbf{w}(a_1 \cdots a_m+)]^2}{N(a_1 \cdots a_m)} \right.$$

$$\left. - \sum_{a_1,\dots,a_{m+1}} \frac{[N_\mathbf{w}(a_1 \cdots a_{m+1})]^2}{N(a_1 \cdots a_{m+1})} + \frac{1 - 2 N_\mathbf{w}(w_1 \cdots w_m+)}{N(w_1 \cdots w_m)} \right].$$

$$(4)$$

In this expression, $N_\mathbf{w}$ means the count inside the word $\mathbf{w}$, $\varepsilon_{\ell-d}(\mathbf{w})$ is equal to 1 if and only if two occurrences of $\mathbf{w}$ can overlap on $\ell - d$ letters and in this case $w_1 \cdots w_d w_1 \cdots w_\ell$ is the resulting composed word ($\varepsilon_{\ell-d}(\mathbf{w}) = 0$ otherwise). The above variable $\widehat{\sigma}_m^2(\mathbf{w})$ corresponds to the quantity sigma2 computed by R'MES.

In the maximal model, namely if $m = \ell - 2$, the formula of the estimated variance reduces to:

$$\widehat{\sigma}_{\ell-2}^2(\mathbf{w}) = \frac{\widehat{\mathbb{E}}_{\ell-2}[N(\mathbf{w})]}{[N(w_2 \cdots w_{\ell-1})]^2} [N(w_2 \cdots w_{\ell-1}) - N(w_2 \cdots w_\ell)]$$

$$\times [N(w_2 \cdots w_{\ell-1}) - N(w_1 \cdots w_{\ell-1})]. \qquad (5)$$

p-**value.** To get the significance of the score $u_m(\mathbf{w})$, one just has to compute the probability:

$$\mathbb{P}(\mathcal{N}(0,1) \geq u_m(\mathbf{w})).$$

High positive values of the score will correspond to exceptionally frequent words (p-value close to 0), whereas negative scores with high absolute values will correspond to exceptionally rare words ($\mathbb{P}(\mathcal{N}(0,1) \leq u_m(\mathbf{w}))$ close to 0).

Generalization to phased models. The score calculated by R'MES for the word $\mathbf{w}$ in phase ϕ is

$$u_{m\_3}(\mathbf{w},\phi) := \frac{N(\mathbf{w},\phi) - \widehat{\mathbb{E}}_{m\_3}[N(\mathbf{w},\phi)]}{\sqrt{\widehat{\sigma}_{m\_3}^2(\mathbf{w},\phi)}} \xrightarrow{\mathcal{D}} \mathcal{N}(0,1),$$

where $\widehat{\sigma}^2_{m\_3}(\mathbf{w}, \phi)$ is derived from $\widehat{\sigma}^2_m(\mathbf{w})$ by adding the relevant phases. As an example, here is the formula in the maximal model with $m = \ell - 2$:

$$
\begin{aligned}
\widehat{\sigma}^2_{m\_3}(\mathbf{w}, \phi) =\ & \frac{\widehat{\mathbb{E}}_{m\_3}[N(\mathbf{w}, \phi)]}{[N(w_2 \cdots w_{\ell-1}, \phi - 1)]^2}[N(w_2 \cdots w_{\ell-1}, \phi - 1) \\
& - N(w_2 \cdots w_\ell, \phi)] \times [N(w_2 \cdots w_{\ell-1}, \phi - 1) \\
& - N(w_1 \cdots w_{\ell-1}, \phi - 1)].
\end{aligned}
\tag{6}
$$

Generalization to word families. The asymptotically Gaussian score for the word family $\mathcal{W}$ is

$$
u_m(\mathcal{W}) := \frac{N(\mathcal{W}) - \widehat{\mathbb{E}}_m[N(\mathcal{W})]}{\sqrt{\widehat{\sigma}^2_m(\mathcal{W})}},
$$

where the estimated variance is the sum of the estimated covariances between all pairs of words in $\mathcal{W}$:

$$
\widehat{\sigma}^2_m(\mathcal{W}) = \sum_{\mathbf{w}, \mathbf{w}' \in \mathcal{W}} \widehat{\sigma}^2_m(\mathbf{w}, \mathbf{w}').
$$

If $\mathbf{w} = \mathbf{w}'$, then $\widehat{\sigma}^2_m(\mathbf{w}, \mathbf{w}) = \widehat{\sigma}^2_m(\mathbf{w})$, and otherwise we have:

$$
\begin{aligned}
\widehat{\sigma}^2_m(\mathbf{w}, \mathbf{w}') =\ & \sum_{d=1}^{\ell-m-1} \varepsilon_{\ell-d}(\mathbf{w}, \mathbf{w}')\widehat{\mathbb{E}}_m[N(w_1 \cdots w_d w'_1 \cdots w'_\ell)] \\
& + \sum_{d=1}^{\ell-m-1} \varepsilon_{\ell-d}(\mathbf{w}', \mathbf{w})\widehat{\mathbb{E}}_m[N(w'_1 \cdots w'_d w_1 \cdots w_\ell)] \\
& + \widehat{\mathbb{E}}_m[N(\mathbf{w})]\widehat{\mathbb{E}}_m[N(\mathbf{w}')] \\
& \times \Bigg[\sum_{a_1, \dots, a_m} \frac{N_{\mathbf{w}}(a_1 \cdots a_m +)N_{\mathbf{w}'}(a_1 \cdots a_m +)}{N(a_1 \cdots a_m)} \\
& \quad - \sum_{a_1, \dots, a_{m+1}} \frac{N_{\mathbf{w}}(a_1 \cdots a_{m+1})N_{\mathbf{w}'}(a_1 \cdots a_{m+1})}{N(a_1 \cdots a_{m+1})} \\
& \quad + \frac{\mathbb{I}\{w_1 \cdots w_m = w'_1 \cdots w'_m\} - N_{\mathbf{w}}(w'_1 \cdots w'_m +)}{N(w'_1 \cdots w'_m)} \\
& \quad - \frac{N_{\mathbf{w}'}(w_1 \cdots w_m +)}{N(w_1 \cdots w_m)} \Bigg],
\end{aligned}
$$

where $\varepsilon_{\ell-d}(\mathbf{w},\mathbf{w}')$ is equal to 1 if and only if an occurrence of $\mathbf{w}$ can be overlapped on $\ell - d$ letters by a later occurrence of $\mathbf{w}'$ ($\varepsilon_{\ell-d}(\mathbf{w},\mathbf{w}') = 0$ otherwise).

Exceptionality score for the skew. The skew of $\mathbf{w}$ is defined by $B(\mathbf{w}) := N(\mathbf{w})/N(\overline{\mathbf{w}})$ where $\overline{\mathbf{w}}$ is the reverse complement (or conjugate) of the word $\mathbf{w}$, for instance gattcg is the conjugate of cgaatc. The p-value to measure the significance of the observed skew $B^{\mathrm{obs}}(\mathbf{w})$ is

$$\mathbb{P}(B(\mathbf{w}) \geq B^{\mathrm{obs}}(\mathbf{w})) = \mathbb{P}(\underbrace{N(\mathbf{w}) - B^{\mathrm{obs}}(\mathbf{w})N(\overline{\mathbf{w}})}_{T(\mathbf{w})} \geq 0)$$

$$= \mathbb{P}\left(\frac{T(\mathbf{w}) - \mathbb{E}[T(\mathbf{w})]}{\sqrt{\mathbb{V}(T(\mathbf{w}))}} \geq -\frac{\mathbb{E}[T(\mathbf{w})]}{\sqrt{\mathbb{V}(T(\mathbf{w}))}}\right).$$

Thanks to the Gaussian approximation of word counts, we can use that $T(\mathbf{w})$ is asymptotically Gaussian with

$$\mathbb{E}[T(\mathbf{w})] = \mathbb{E}[N(\mathbf{w})] - B^{\mathrm{obs}}(\mathbf{w})\mathbb{E}[N(\overline{\mathbf{w}})]$$

$$\mathbb{V}(T(\mathbf{w})) = \mathbb{V}[N(\mathbf{w})] - 2B^{\mathrm{obs}}(\mathbf{w})\mathbb{C}\mathrm{ov}(N(\mathbf{w}), N(\overline{\mathbf{w}}))$$

$$+ [B^{\mathrm{obs}}(\mathbf{w})]^2\mathbb{V}[N(\overline{\mathbf{w}})]$$

and that $\frac{T(\mathbf{w})-\mathbb{E}[T(\mathbf{w})]}{\sqrt{\mathbb{V}(T(\mathbf{w}))}}$ is asymptotically distributed like a $\mathcal{N}(0,1)$. Therefore, we can derive the following score

$$\frac{-\widehat{\mathbb{E}}_m[N(\mathbf{w})] + B^{\mathrm{obs}}(\mathbf{w})\widehat{\mathbb{E}}_m[N(\overline{\mathbf{w}})]}{\sqrt{\widehat{\sigma}_m^2(\mathbf{w}) - 2B^{\mathrm{obs}}(\mathbf{w})\widehat{\sigma}_m^2(\mathbf{w},\overline{\mathbf{w}}) + [B^{\mathrm{obs}}(\mathbf{w})]^2.\widehat{\sigma}_m^2(\overline{\mathbf{w}})}}.$$

5.4. Clumping occurrences

This section will be crucial for the compound Poisson approximation of the count. It relies on the fact that occurrences of a given word may overlap in a sequence and that occurrences will occur in clumps. A clump of $\mathbf{w}$ is a maximal set of overlapping occurrences of $\mathbf{w}$ in a sequence. By definition, clumps do not overlap each other and the number of occurrences of $\mathbf{w}$ in a clump of $\mathbf{w}$ is called the *size* of the clump. Actually, only *periodic* (or *overlapping*) words will tend to occur in clumps.

Periods of a word. An integer $p \in \{1, \ldots, \ell - 1\}$ is said to be a period of $\mathbf{w}$ if and only if two occurrences of $\mathbf{w}$ can start at a distance p apart; we denote by $\mathcal{P}(\mathbf{w})$ the set of periods of the word $\mathbf{w}$. In other words,

$$p \in \mathcal{P}(\mathbf{w}) \Longleftrightarrow \varepsilon_{\ell-p}(\mathbf{w}) = 1$$

$$\Longleftrightarrow w_j = w_{j+p} \quad \text{for all } j \in \{1, \ldots, \ell - p\}.$$

For instance $\mathcal{P}(\texttt{aataataa}) = \{3, 6, 7\}$. A *nonoverlapping* word is a word which has no period, for instance, $\mathcal{P}(\texttt{aatcc}) = \emptyset$.

Periods that are not a strict multiple of the smallest period are said to be *principal* since they will be more important, as we will see later. $\mathcal{P}'(\mathbf{w})$ denotes the set of the principal periods of $\mathbf{w}$, for instance $\mathcal{P}'(\texttt{aataataa}) = \{3, 7\}$.

5.5. *Compound Poisson approximation*

The asymptotic normality of the count distribution requires that the expected count tends to infinity with the length of the sequence. In practice (fixed sequence length), it means that the expected count should be large enough so that the limiting Gaussian distribution is not truncated at zero. When the expected count is too small, it means that the occurrences are rare and they can be considered like a compound Poisson process along the sequence. Indeed, the clumps are asymptotically Poisson distributed and the compound feature is due to the clump size (Schbath, 1995). The key formulas are the following:

$$N(\mathbf{w}) = \sum_{c=1}^{\widetilde{N}(\mathbf{w})} K_c(\mathbf{w}) = \sum_{k=1}^{\infty} k \widetilde{N}_k(\mathbf{w}),$$

where $\widetilde{N}(\mathbf{w})$ denotes the number of clumps of $\mathbf{w}$, $\widetilde{N}_k(\mathbf{w})$ denotes the number of clumps of sike k and $K_c(\mathbf{w})$ represents the size of the c-th clump.

Limiting Poisson distribution. It has been shown that if $\mathbf{w}$ is nonoverlapping or has no period less or equal to $\ell - m$, then the count $N(\mathbf{w})$ can be approximated by a Poisson random variable with mean $\mathbb{E}_m[N(\mathbf{w})]$. In that case, the p-value $\mathbb{P}(N(\mathbf{w}) \geq N^{\text{obs}}(\mathbf{w}))$ is just the upper tail of the Poisson distribution $\mathcal{P}(\widehat{\mathbb{E}}_m[N(\mathbf{w})])$.

Limiting compound Poisson distribution. When $\mathbf{w}$ has periods less or equal to $\ell - m$, then the number of clumps $\widetilde{N}(\mathbf{w})$ can be approximated by a Poisson random variable with mean $(1 - a_m(\mathbf{w}))\mathbb{E}_m[N(\mathbf{w})]$ where

$$a_m(\mathbf{w}) = \sum_{p \in \mathcal{P}'(\mathbf{w}),\, p \le \ell - m} \prod_{j=1}^{p} \pi(w_j \cdots w_{j+m-1}, w_{j+m}) \qquad (7)$$

and the clump size is asymptotically geometrically distributed:

$$\mathbb{P}(K_c(\mathbf{w}) = k) = (1 - a_m(\mathbf{w}))[a_m(\mathbf{w})]^{k-1}.$$

The quantity $a_m(\mathbf{w})$ can be interpreted like the overlapping probability of $\mathbf{w}$. A compound Poisson distribution is the distribution of $\sum_{c=1}^{Z} K_c$ where Z is a Poisson random variable and the K_c's are independent and identically distributed. In the particular case where the K_c's follow the geometric distribution with parameter a, the associated compound Poisson distribution is called Geometric-Poisson or Pólya-Aeppli distribution (Johnson et al., 1992), and the parameters are $(\mathbb{E}Z, a)$.

The p-value $\mathbb{P}(N(\mathbf{w}) \ge N^{\text{obs}}(\mathbf{w}))$ is then the upper tail of the Geometric-Poisson distribution with parameters $((1 - a_m(\mathbf{w}))\mathbb{E}_m [N(\mathbf{w})], a_m)$. The algorithm from Nuel (2008) is used to compute such upper tail. Both quantities a_m and $(1 - a_m(\mathbf{w}))\mathbb{E}_m[N(\mathbf{w})]$ are calculated by R'MES under the names A and expect_p in the output tables.

Generalization to word families. When studying the count of a word family $\mathcal{W}$, one has to consider possible overlaps between any two words from the family and to work with the clumps composed of overlapping occurrences of $\mathcal{W}$. It results from Roquain and Schbath (2007) that $N(\mathcal{W})$ can be approximated by a general compound Poisson distribution whose parameters are derived from an overlapping probability matrix $(a_m(\mathbf{w}, \mathbf{w}'))_{\mathbf{w},\mathbf{w}' \in \mathcal{W}}$. The p-value is obtained by summing up all the point probabilities of a compound Poisson distribution calculated thanks to the algorithm from Barbour et al. (1992b).

References

Barbour AD, Holst L, Janson S. (1992b) *Poisson Approximation*. Oxford University Press.

Bigot S, Saleh O, Lesterlin C et al. (2005) KOPS: DNA motifs that control E. coli chromosome segregation by orienting the FtsK translocase. *EMBO J* **24**: 3770–3780.

Chiapello H, Bourgait I, Sourivong F et al. (2005) Systematic determination of the MOSAIC structure — backbone versus strain specific loops — of bacterial genomes. *BMC Bioinformatics* **6**: 171.

Halpern D, Chiapello H, Schbath S *et al.* (2007) Identification of DNA motifs implicated in maintenance of bacterial core genomes by predictive modelling. *PLoS Genetics* **3**: e153.

Johnson NL, Kotz S, Kemp AW. (1992) *Univariate Discrete Distributions*, Wiley, New York.

Karlin S, Burge C, Campbell AM. (1992) Statistical analyses of counts and distributions of restriction sites in DNA sequences. *Nucleic Acids Res* **20**: 1363–1370.

El Karoui M, Biaudet V, Schbath S, Gruss A. (1999) Characteristics of Chi distribution on several bacterial genomes. *Research in Microbiology* **150**: 579–587.

Lothaire M. (2005) *Applied Combinatorics on Words*, vol. 105 of *Encyclopedia of Mathematics and its Applications*. Cambridge University Press.

Nuel G. (2008) Cumulative distribution function of a geometric Poisson distribution. *J Stat Comput Simulat* **78**: 385–394.

Prum B, Rodolphe F, de Turckheim E. (1995) Finding words with unexpected frequencies in deoxyribonucleic acid sequences. *J R Stat Soc B* **57**: 205–220.

Reinert G, Schbath S, Waterman M. (2000) Probabilistic and statistical properties of words. *J Comput Biol* **7**: 1–46.

Robin S, Rodolphe F, Schbath S. (2005) *DNA, Words and Models*. Cambridge University Press, English version of *ADN, mots et modèles*, BELIN 2003.

Robin S, Schbath S. (2001) Numerical comparison of several approximations of the word count distribution in random sequences. *J Comput Biol* **8**: 349–359.

Robin S, Schbath S, Vandewalle V. (2007) Statistical tests to compare motif count exceptionalities, *BMC Bioinformatics* **8**: 1–20.

Roquain E, Schbath S. (2007) Improved compound Poisson approximation for the number of occurrences of multiple words in a stationary Markov chain. *Advances in Applied Probability* **39**: 128–140.

Schbath S. (1995) Compound Poisson approximation of word counts in DNA sequences. *ESAIM: Probability and Statistics* **1**: 1–16.

Smith HO, Tomb JF, Dougherty BA, Fleischmann RD, Venter JC. (1995) Frequency and distribution of DNA uptake signal sequences in the Haemophilus influenzae Rd genome. *Science* **269**: 538–540.

Sourice S, Biaudet V, El Karoui M, Ehrlich SD, Gruss A. (1998) Identification of the Chi site of *Haemophilus influenzae* as several sequences related to the *Escherichia coli* Chi site. *Mol Microbiol* **27**: 1021–1029.

Chapter 3

An Intricate Mosaic of Genomic Patterns at Mid-range Scale

Alexei Fedorov and Larisa Fedorova*

1. Introduction

Genomic patterns on short-range scales represent various "words" composed from nucleotide "letters." Each of these words occurs many times within DNA sequences. The longest words, also known as "pyknons," are up to 17-nucleotide-long sequences which are over-abundant in the exons and introns of humans and other mammals (Rigoutsos *et al.*, 2006; Tsirigos and Rigoutsos, 2008). The vast majority of sequences only a little bit longer than pyknons are unique even for the large genomes of animals and plants. For example, the complete theoretical set of 20-nucleotide-long sequences is comprised of 4^{20} different words of length 20, which is just over one trillion. More than 99% of these 20-mer oligonucleotides never occur in the entire human genome ($\sim 3 \times 10^9$ bp). Therefore, biologists frequently use 20-mer oligonucleotides as PCR primers or hybridization probes for experimental characterization of particular genomic segments. The genomic arrangement of short sequences (<20 bp) is covered in other chapters of this book. Here we consider genomic patterns longer than 30 and up to several thousands of nucleotides to be called mid-range scale. At this mid-range, most of the sequences are unique, i.e. occur only once in the entire genome, hence, it is more appropriate to characterize or group them not by their exact sequence of nucleotides but rather by their overall nucleotide composition, such as (G+C)-richness, purine-richness, etc. We also distinguish mid-range genomic scales from the long-range one

---

*Department of Medicine, The University of Toledo, Health Science Campus, Ohio, USA

represented by genomic isochores reviewed elsewhere (Bernardi, 2007). Traditionally, (G+C)-rich and (G+C)-poor isochores are considered to be from 100 kb and longer. Recently, scientists have started to describe ultra-short isochores in the range of tens of thousands of nucleotides. In order not to interfere with isochores, we limit the length of mid-range patterns by ten thousand bases. The main focus of this chapter is to show that at mid-range scales, genomes of complex eukaryotes consist of a number of different patterns and are associated with unusual DNA conformations. Some of these patterns are scarcely investigated and still waiting for thorough exploration and recognition.

2. Results and Discussion

2.1. *DNA repeats — important elements at genomic mid-range scale*

All eukaryotic genomes contain several extra-large "words" recurring many times — so called DNA repetitive elements, the size of which are generally within mid-range scale. DNA repeats are classified into three major classes based on the molecular mechanisms of their origin and propagation: transposons, retrotransposons, and tandemly organized repeats. There is a large variety of transposons and retrotransposons that can be specific for narrow taxons (like the Alu-elements within primates), or that have a much broader representation (like the L1-repeats found in all vertebrates). We will not consider these DNA repetitive elements, but only refer a reader to several excellent, detailed reviews on their genomic organization and evolution (Jurka *et al.*, 2007; Eickbush and Jamburuthugoda, 2008; Richard *et al.*, 2008). Here we concentrate only on the simple tandem repeats that exist in almost every eukaryotic organism. Our examples well illustrate the common trend for mid-range scale sequence patterns to associate with DNA conformation abnormalities or alternative 3D structures.

We begin the examination of a simple tandem repeat from a well-characterized type composed of a reiterating pentamer sequence AATGG. Our computer analysis of completely sequenced mammalian genomes demonstrates that there were 162 different loci inside the euchromatic genomic regions of humans, 24 in mouse, 14 in rat, 21 in cow, and 58 in dog that contain $(AATGG)_N$ perfect repeats, where $N \geq 4$. These sequences are proportionally distributed between intergenic regions and introns and, often, there are up to several dozens of tandem AATGG pentamers in one locus. The location of these repeats inside genes is not evolutionarily conserved, since we have not detected their presence in the same intron of the named species. In addition, the same tandemly repeated pentamer AATGG is one of the most evolutionarily conserved parts of a centromere, where it exists in thousands of copies and serves as an attachment point for the two sister chromosomes during mitosis (Grady *et al.*, 1992; Lee *et al.*, 1997). Interestingly, under physiological conditions, this DNA-repeat comprised of at least four pentamers could exist not only as B-form Watson–Crick duplex but also in an unusual form with highly asymmetrical conformations of AATGG-strand and its complementary CCATT-strand (Jaishree and Wang, 1994; Catasti *et al.*, 1999). Its transition from Watson–Crick duplex to single-stranded structures is facilitated by acidic pH conditions. Figure 1 demonstrates the NMR solution structure of the anti-parallel stranded non-B-form DNA duplex 5'-TGGAATGGAA:TGGAATGGAA-3' created by two repetitive pentamers published by Chou *et al.* (1994). This particular structure is also known as "interdigitated" or zipper-like stacking (Chou *et al.*, 2003). The scheme of the unusual 3D structure of AATGG repeats is illustrated in Figs. 8–10 of Catasti and co-authors (Catasti *et al.*, 1999) while a slightly different variant of spatial organization of the same repeat is illustrated in Fig. 1 of Jaishree and Wang (1994). These two papers demonstrate that the AATGG strand of the repeat forms stable doubly folded hairpins with Watson–Crick A-T and non-Watson–Crick A-G and G-G base pairs in the stems. The stability of these stems is reached partially due to stacking of the three purines shown by arrows in Fig. 1. Moreover, the same authors demonstrated that a greater number of the AATGG pentamers might form higher-order structure in which doubly folded hairpins are compactly organized in a helical array of $(AATGG)_4$ units. At the same time, the complementary strand formed by CCATT pentamers is unstable under physiological conditions and likely represents loose structures. Under acidic conditions, this CCATT tandem repeat might also form unusual structures known as i-motif with intercalated cytosine bases shown in Fig. 2 (Catasti *et al.*, 1999; Nonin-Lecomte and Leroy, 2001).

Other tandemly organized simple repeats could also have conformations far distinct from Watson–Crick double helices. One of the prominent noncanonical structures, known as G-quadruplex, G-quartet, or

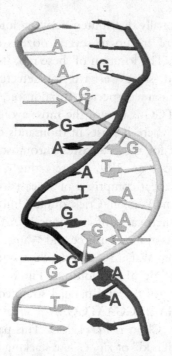

Fig. 1. Cartoon of 3D-structure of anti-parallel DNA duplex 5'TGGAATGGAA: TGGAATGGAA3' formed by two copies of TGGAA pentamer repeat. This picture is a snapshot of the structure with the identifier 103D obtained from the Protein Data Bank. The structure was resolved via solution NMR approach by Chow and co-authors (Chou *et al.*, 1994). Four arrows point to unpaired guanosine residues stacked between Hoogsteen G-A pairs.

G4, is formed by guanine-rich strands of the repeats. Quadruplexes are arranged in four-stranded structures with stands connected to each other via Hoogsteen hydrogen bonding. G-quadruplex has been well characterized in human telomeric and related sequences with the core repetitive element TTAGGGG and also within promoters and 5'-untranslated regions of human genes whose sequences have a loose consensus of $G_{3-5}N_{L1}G_{3-5}N_{L2}G_{3-5}N_{L3}G_{3-5}$, where N_{L1}, N_{L2}, and N_{L3} are loops with the length from 1 to 7 nucleotides and variable nucleotide composition (Neidle, 2009). There are several alternative conformations of G-quadruplexes due to the organization of the strands relative to each other. Among them are anti-parallel, parallel, and parallel/anti-parallel hybrids (Oganesian and Bryan, 2007; Huppert, 2008). One example of a

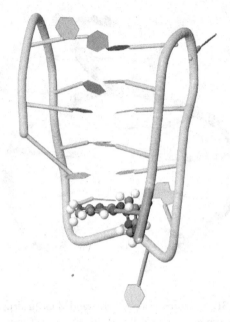

Fig. 2. Cartoon of 3D-structure of a C-rich strand fragment of the human centromeric satellite III d(CCATTCCATTCCTTTCC) that forms intramolecurlar i-motif structure with C.C(+) pairs from parallel strands intercalated head-to-tail. This picture is a snapshot of the structure with the identifier 1G22 obtained from the Protein Data Bank. The structure was resolved via solution NMR approach by Nonin-Lecomte and Leroy (2001) for uridine derivative methylated on the first cytidine base, d(5mCCATTCCAUTCCUTTCC), whose proton spectrum is better resolved. Modified residue 5-METHYL-2′-DEOXYCYTIDINE is demonstrated with white, blue, red, and grey spheres.

parallel-stranded G4 NMR structure is illustrated in Fig. 3. G-quadruplex structure has been demonstrated for several G-rich short tandem repeats. Among them are GGA and CGG triplet repeats ((Matsugami *et al.*, 2003; Nakagama *et al.*, 2006) and the GGCAG mouse minisatellite Pc-1 (Katahira *et al.*, 1999). Four elements of GGA-repeat can form intramolecular parallel quadruplex, while the neighboring quadruplexes can form a dimer stabilized through the stacking interaction between the heptads of the two quadruplexes (Matsugami *et al.*, 2003). While G-quadruplexes are formed by a guanine-rich strand, their complementary strand being C-rich may also form a completely different four-stranded structure known as i-motif or intercalated cytosine tetraplex. This structure is only stable

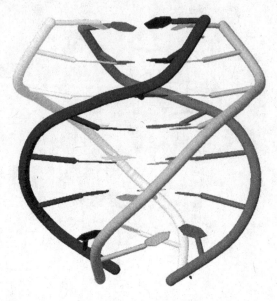

Fig. 3. Cartoon of 3D-structure of a parallel-stranded G-quadruplex DNA formed by the Tetrahymena telomeric sequence d(T-T-G-G-G-G-T). This picture is a snapshot of the structure with the identifier 139D obtained from the Protein Data Bank. The structure was resolved by combined NMR-computational approach by (Wang and Patel, 1993).

under acidic conditions (Huppert, 2008). Intercalated cytosines have been found in several unusual conformations, one of which is shown in Fig. 2.

There are up to 10 different non-B-from DNA conformations associated with the simple repeats listed and well illustrated by Wells (Wells, 2007). Among them are slipped structure formed by CNG repeats; triplex DNA formed by purine (R)-rich or pyrimidine (Y)-rich mirror repeats (described below in details); sticky DNA formed by (G+A)-rich tracts like $(GAA)_N$; and DNA unwinding elements formed by (A+T)-rich regions. In addition, Wells describes cruciforms created by inverted repeats; and left-handed Z-DNA formed by alternated R/Y bases in $(RY)_N$ repeats (Wells, 2007). There is evidence that CGG triplet repeats could form a non-B-type higher order structure (Nakagama *et al.*, 2006). This particular CGG repeat is widespread in animal genomes and it expands inside the first exon of the FMRI gene that causes Fragile X syndrome (Mandel, 1993; Crawford *et al.*, 2001). NMR and X-ray crystallography studies of DNA oligonucleotides strongly suggest that $(GA)_N$

and $(A)_N$ repeats could form parallel-stranded homeoduplexes (Kypr *et al.*, 2007; Chakraborty *et al.*, 2009). However it is questionable whether such structures might exist *in vivo*.

Tandem repeats with longer units could also form non-B structures. For instance, the $(ACAGGGGTGTGGGG)_N$ insulin minisatellite has a complex loop-folding conformation (Catasti *et al.*, 1999). The listed non-B DNA conformations likely do not exist permanently, but only under specific conditions. Their formation can be facilitated by negative supercoiling during transcription or by binding with transcription factors (Mirkin, 2008). The very specific pattern of mutagenesis within simple repeats associated with particular bases and particular sites strongly suggests the existence of non-B structures *in vivo* (Wells, 2007). Also, several non-B structures have been confirmed in various experiments including *in vivo* studies (Wells, 2007; Fernando *et al.*, 2009; Kypr *et al.*, 2009). Currently, more than 70 human genetic disorders have been associated with changes in simple repeats (Lupski, 1998; Wells, 2007).

In summary, simple repeats are abundant in the genomes of diverse animals and plants. In rodents, 2.4% of the euchromatic part of their genome is represented by simple repeats, which is two times bigger than the length of all protein-coding sequences (Gibbs *et al.*, 2004). Additionally, tandemly organized short sequences are abundant and are key components of telomeres and centromeric regions. Many simple repeats, whose total length reaches 20 bases and above, under certain conditions can exist in a variety of non-B DNA conformations *in vivo* associated with specific genomic functions. Computationally, simple repeats can be detected by the RepeatMasker program (Smit AFA). However, the default parameters of this program could skip recognition of simple repeat loci whose copy numbers are low or where the sequences have accumulated mutations (fuzzy repeats). In this case the best choice is the stand-alone Tandem Repeat Finder with advanced search parameters (Benson, 1999).

2.2. Genomic Mid-Range Inhomogeneity (MRI): Nucleotide compositional extremes and sequence nonrandomness

In thousands of genomic regions, the composition of A, T, C, or G content or different combinations of these bases exist at extremes far different from the average base composition. We call such compositional extremes genomic mid-range inhomogeneity or MRI if they stretch at least 30 base pairs but less than 10 000 base pairs. To characterize genomic MRI patterns, a public computational resource (*Genomic MRI*) has been created that allows detecting sequence regions with any type of extreme composition (Bechtel *et al.*, 2008). Using this resource it was demonstrated that various MRI regions occupy up to a quarter of the human genome and their existence is maintained via strong fixation bias (Prakash *et al.*, 2009).

2.2.1. *Genomic MRI toolkit*

For examining mid-range sequence patterns, *Genomic MRI* programs do not characterize particular "words" but only the overall compositional content of particular base(s) that we refer to as X (X could be a single nucleotide A, G, C, or T or any of their combinations like A+C, or G+T+C, etc.). *Genomic MRI* allows studying the distribution of X-rich regions in any sequence of interest. These X-rich MRI regions are highly over-represented in mammalian genomes for all kinds of X-contexts. For instance, in the human genome, (G+C)-rich sequences with lengths from 100 to 200 nucleotides are 20 times over-represented; (A+T)-rich sequences in the same length range are about 12 times over-represented; (A+G)-rich and (T+C)-rich sequences 10 times; and (G+T)-rich and (A+C)-rich sequences up to six times over-represented (Bechtel *et al.*, 2008). In order to measure the abundance of X-rich regions in the sequences under analysis, *Genomic MRI* compares their presence inside a specifically generated random sequence that has the same oligonucleotide distribution as the real one. This evaluation is achieved by the following computational steps. Firstly, the short-range inhomogeneity (SRI) of a given sequence is analyzed by the SRI-*analyzer* program from the *Genomic MRI* package to create an oligonucleotide frequency table for each possible 1–9 nucleotide long "word." Then, a second program, SRI-*generator*, creates a random sequence with a short-range inhomogeneity that approximates the oligonucleotide frequency table of the natural sequence. This random sequence is used further for comparison with the natural one. Finally, the third program, MRI-*analyzer*, scans a sequence under analysis and the random sequence with a window of a specified size and checks whether the nucleotide composition of the sequence in the current window is X-rich or X-poor for a particular chosen combination of nucleotides (X), e.g. A, T, C, G, G+C, A+G, G+T, etc. A window is rich for the X-content if its X-composition is above a user-specified threshold-X_1, while a window is X-poor if it is below another user-specified threshold-X_2. (Note that X-poor regions can be referred to as non-X-rich ones, e.g. (G+C)-poor are (A+T)-rich). An example of *MRI analyzer* graphical output is shown in Fig. 4 that illustrates the

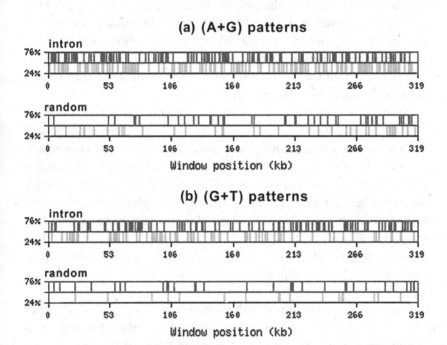

Fig. 4. The graphical output of the *MRI-analyzer* program for the first intron of the Dystrophin gene (marked as "intron") and also for the SRI-generated random sequence based on the tetramer oligonucleotide frequency table of the intron (marked as "random"). The entire sequence of the 319 kb intron and the random sequence is displayed on the *x*-axis. Dark gray bars on each top row represent positions of content-rich MRI regions on the sequence. Light gray bars on the bottom row represent content-poor MRI regions. The *y*-axis contains upper and lower thresholds for the given content type. (a) *Genomic MRI* analysis of (A+G)-rich and (A+G)-poor (or (T+C)-rich) regions; (b) *Genomic MRI* analysis of (G+T)-rich and (G+T)-poor (or (C+A)-rich) regions.

MRI-patterns for an extra-large human intron of the dystrophin gene from chromosome X.

Two scales of MRI regions should be considered. First, regions from 30 to 1000 bp, whose properties have been investigated in detail and for which several periodicities have been reported (Trifonov, 1991; Herzel *et al.*, 1999; Ioshikhes *et al.*, 1999). Second, larger regions from 1 to 10 kb, which are one of the least studied areas in genomic composition and where as yet unknown biological properties may be found. Such sub-divisions are important for the proper choice of parameters for the MRI thresholds. For instance, for a 100-nucleotide-long window, there are a vast number of regions in mammals where (G+C) composition is 85% or higher. However, for studying regions with a window-size of around 5 kb, the upper threshold for (G+C) content should not be more than 65% to find the areas satisfying the criterion.

Extended regions with compositional extremes satisfying (G+C)- or (A+T)-richness are abundant in vertebrates and can be as long as several million bases (known as genomic isochores). Other composition extremes, such as R-, Y-, (G+T)-, (A+C)-richness, that extend over long chromo-somal regions are not as abundant as (C+G)- and (A+T)-rich genomic areas. Nonetheless, for more than 2100 human chromosomal regions with lengths exceeding 10 kb, we have detected frequencies of more than 60% for (G+A)-, (T+C)-, (A+C)-, or (G+T)-nucleotides. As for extremes, our computations have shown that there are 22 regions in the human genome where R or Y composition exceeds 70% within a sequence longer than 10 kb.

Recently, by studying the distribution of more than four million SNPs in the human genome and by taking into account their frequencies in the population, the influence of mutations on different MRI regions has been examined (Prakash *et al.*, 2009). The authors demonstrated that MRI regions have comparable levels of *de novo* mutations to the control genomic sequences with average base composition. *De novo* substitutions rapidly erode MRI regions, bringing their nucleotide composition toward genome-average levels. However, those substitutions that favor the main-tenance of MRI properties have a higher chance to spread through the entire population. All in all, the observed strong fixation bias for muta-tions helps to preserve MRI regions during evolution, indicating their involvement in genomic operations.

2.2.2. (G+C)-rich and (A+T)-rich MRI regions are associated with several unusual DNA structures

We start considering mid-range genomic compositional patterns from the most studied case: (G+C)-rich and (A+T)-rich regions. These (G+C)-rich and (A+T)-rich regions of various lengths from thirty to several thousand nucleotides are 4–20 times over-represented in the mammalian genomes compared to random expectation (Bechtel, 2008; Bechtel *et al.*, 2008). Among (G+C)-rich genomic segments, CpG-islands have drawn the most public attention, due to their functional properties and involvement in gene expression regulation (Hackenberg *et al.*, 2006). CpG-islands are found in nearly 60% of human genes including almost all of the house-keeping ones (Hackenberg *et al.*, 2006). According to two different definitions of these islands, their length must be at least 200 or 500 bp long; (G+C) content more than 50 or 55%; and the number of CpG dinucleotides in the islands should exceed more than twice their occurrence in other genomic regions (Gardiner-Garden and Frommer, 1987; Takai and Jones, 2003; respectively). CpG dinucleotides are important sites for cytosine methylation in all vertebrates and some invertebrates and plants. However, inside CpG-islands CpG dinucleotides are predominantly nonmethylated (Suzuki and Bird, 2008). It has been shown recently that CpG dinucleotide without methylation exhibit structural abnormalities in the DNA helix. Particularly, they are one of the most frequent sites for DNA backbone cleavage by hydroxyl radicals (Greenbaum, Pang, and Tullius, 2007; Greenbaum, Parker, and Tullius, 2007) and during the sonication of double-stranded DNA (Grokhovsky *et al.*, 2008). The crucial involvement of cytosine methylation in the regulation of gene expression is well described in a number of reviews including some recent ones (Prokhortchouk and Defossez, 2008; Suzuki and Bird, 2008; Illingworth and Bird, 2009). Thus, here we concentrate on the other physicochemical properties of (G+C)-rich and (A+T)-rich regions.

It is well known that A-form of DNA helix exists in high salt concentrations and in ethanol-containing solutions. However, (G+C)-rich regions may be present in A-form DNA even in aqueous solutions (Warne and deHaseth, 1993; Stefl *et al.*, 2001; Kypr *et al.*, 2009). A special form of DNA which is an intermediate between A- and B-forms, has been characterized in (G+C)-rich sequences with methylated cytosines (Vargason *et al.*, 2000). In addition, short $(CpG)_n$ repeats could adopt Z-DNA as recently reviewed by P.S. Ho (Ho, 2009). This Z-DNA is proposed to serve as transcriptional co-activator (Liu *et al.*, 2001).

(A+T)-rich regions, on the other hand, are also associated with special DNA conformations. Some of these sequences with specific distributions of A and T bases form an unusual structure known as the DNA unwinding element (Kowalski *et al.*, 1988). These elements are often associated with the origins of replication in eukaryotes and prokaryotes (Umek *et al.*, 1989). There are several (A+T)-rich simple repeats widespread in eukaryotes. Among them, $(AT)_n$ is one of the most common in animals. X-ray and NMR studies of the DNA oligomer d(ATATAT) have shown that in addition to B-DNA, it could form an anti-parallel double helical duplex in which the base pairing is of the Hoogsteen type (Abrescia *et al.*, 2004). The adenines in this duplex are flipped over making the minor groove narrow and hydrophobic. This structure is very similar to the standard B-form helix with about 10 base pairs per turn. Theoretical analysis has demonstrated that energies of the Hoogsteen form and B-form of DNA are practically identical (Cubero *et al.*, 2003). Most recently, Chakraborty and co-authors demonstrated that poly-dA oligonucleotides (dA_{15}) under acidic pH conditions could allow the formation of a double-helical parallel-stranded duplex held together by reversed Hoogsteen type AH^+-H^+A base pairs (Chakraborty *et al.*, 2009).

(A+T)-rich regions presumably have several important cellular functions. First, the most indicative compositional characteristic of scaffold/matrix-attached regions is that they are (A+T)-rich (Liebich *et al.*, 2002). Second, centromere DNA of diverse animals, plants and fungi always contain (A+T)-rich regions (Choo, 1997; Abrescia *et al.*, 2004).

2.2.3. *R-rich/Y-rich MRI regions are associated with H-DNA triplex*

All combinations of nucleotide pairs except (G+C) and (A+T) have strand asymmetry. For example, if one strand is enriched by purines (R), the complementary strand is enriched by pyrimidines (Y). Therefore, R- and Y-rich sequences and also (T+G)- and (A+C)-rich ones have physically the same loci, yet representing complementary strands. From here on we will consider them together and refer to them as R/Y-rich and (T+G/A+C)-rich, respectively.

Since 1957, it has been shown that complementary DNA strands, one of which is R-rich and another Y-rich, can form three-stranded helical structures or triplexes (Felsenfeld and Rich, 1957). Intramolecular triplexes, known also as H-DNA, materialize under certain conditions, like supercoiling, when half of the DNA duplex may dissociate into single strands and one of the stand-alone strands can interact via Hoogsteen base pairing with the remaining Watson–Crick DNA duplex along its major groove forming a triplex structure. The remaining stand-alone strand stays unpaired. An example of a DNA triplex is shown in Fig. 5. There are four kinds of H-DNA depending on strand type and orientation (Jain *et al.*, 2008). One type of H-DNA forms under acidic conditions when the stand-alone Y-rich strand interacts with the R-rich strand of the remaining duplex. Particularly, thymines of the stand-alone strand interact with

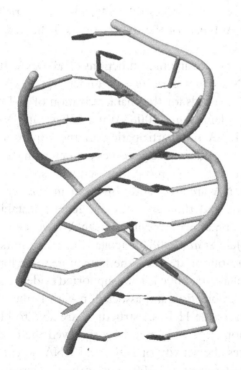

Fig. 5. Cartoon of 3D-structure of a purine.purine.pyrimidine DNA triplex containing G.GC and T.AT triples. This picture is a snapshot of the structure with the identifier 134D obtained from the Protein Data Bank. The structure was resolved using a combined NMR and molecular dynamics approach by Radhakrishnan and Patel (1993).

adenosines of the A-T Watson–Crick pairs of the duplex via Hoogsteen hydrogen bonding, while cytosines of the stand-alone strand interact with guanines of G-C Watson–Crick pairs. Due to this base match requirement for the assembly of this kind of triplex, the sequences of Y-rich stand-alone strand and the Y-rich strand in the duplex should have sequence mirror symmetry. (Here is an example of two sequences with mirror symmetry: 5′-TAGTTCC-3′ and 5′-CCTTGAT-3′.) In many R/Y-rich regions of the genomes, such mirror symmetry has been observed. For example, a 2.5 kb R-rich sequence of the 21st intron of the human PKD1 gene has 23 mirror repeats that form H-DNA (Van Raay et al., 1996; Blaszak et al., 1999). Another kind of intramolecular triplex, can be formed at neutral pH and requires bivalent cations for stability. It is formed by the interaction of R-rich stand-alone strand with the remaining duplex via Hoogsteen bonding. It does not require strong mirror symmetry within its sequences, since the adenines of the stand-alone R-rich strand could interact with the A-T pair of the duplex or with the G-C pair (Malkov et al., 1993).

There are several documented functions of H-DNA. It is well established that H-DNA could exist *in vivo* under certain conditions. Various experimental methods for the characterization of H-DNA have been reviewed recently (Jain et al., 2008; Wang, Zhao, and Vasquez 2009). Single stranded DNA not participating in the triplex is accessible to S1-nuclease cleavage. Eukaryotic genomes contain many S1-nuclease sensitive sites within runs of homo-purine sequences. These segments of single-stranded DNA are frequently involved in the recombination of homologous DNA and thus are sites for genetic instability. Different schemes of recombination involving H-DNA have been described by Jain and others (Jain et al., 2008). Bacolla with co-authors characterized nearly 3000 homo-purine tracks in the human genome longer than 100 nucleotides (Bacolla et al., 2006). They supported evidence for these tracks in promoting recombination and association with higher rates of mutations. In addition, stable H-DNA structures are able to block transcription and replication. Jain and co-authors surveyed the evidence for how H-DNA influences the activity of DNA and RNA polymerases. Finally, Goni and others (Goni et al., 2006) performed a large-scale bioinformatic analysis of the distribution of short R-rich sequences in the human genome. They demonstrated that short R-rich sequences are several times more abundant in the downstream promoter regions compared to other regions and to random expectation models. These short R-rich sequences hold evolutionary conservation between human and mouse yet; likely they are not direct targets for transcription factors. Goni and co-authors have suggested that these sequences act as pacing fragments in promoter regions and help in the correct positioning of transcription factors.

2.2.4. DNA and RNA properties of GT-rich/AC-rich MRI regions

Recall that the complementary strands of (G+T)-rich regions are naturally (A+C)-rich regions. They co-exist with each other and we consider them interchangeably with respect to their description in the literature. According to nucleic acid nomenclature, G or T nucleotides are also known as *Keto* or K while A or C are known as *aMino* or M (Moss). Thus, sometimes these regions are referred to as K.M-tracks or motifs (Yagil, 2004). Bechtel and co-authors demonstrated that (G+T)-regions are about five times more abundant in the mammalian genomes compared to random expectation (Bechtel *et al.*, 2008). Moreover, these regions practically do not intersect with interspersed DNA repeats at all. In 2004 Yagil demonstrated that K.M motifs are significantly over-represented in the genomes of diverse animals, plants, and fungi. Specifically, K.M motifs are predominant in the *D. melanogaster* genome, where they outnumber other motifs such as R/Y-rich motifs (Yagil, 2004). Despite their abundance, (G+T)-rich motifs are much less investigated than other regions with extremes in base compositions. Possible functions that could be associated to (G+T)-rich regions are the following. Firstly, $(CA)_N$ simple repeats are one of the most profuse tandem repeats in mammalian genomes (Waterston *et al.*, 2002). They also should be considered as alternating R/Y sequence, and, due to this property, associated with a Z-DNA conformation (Vogt *et al.*, 1988), which is considered in the next section. Second, C-rich regions, which could be a component of CA-rich regions, are capable of forming four-stranded intercalated molecules (Berger *et al.*, 1996). We mentioned such structures (i-motifs) above in the Simple Repeat section and present an example of it in Fig. 2. Third, telomeres of various eukaryotic species are represented by (G+T)-rich regions which form G-quadruplexes (see above). Fourth, short (G+T)-rich regions could represent transcription factor binding sites such as for factor Sp1 (Wang *et al.*, 2009). Intriguingly, (G+T)-rich oligonucleotides possess antiviral activities. For example, $T_2(G_4T_2)_3$ sequences are virucidal against herpes simplex virus (Shogan *et al.*, 2006). At the RNA level, (C+A)-rich sequences within intronic segments could regulate alternative splicing by being binding sites for the hnRNP L protein (Hui *et al.*, 2005). The presence of (C+A)-rich sequences at the 3'-UTR of mRNA could regulate gene expression at the level of translation (Hamilton *et al.*, 2008). The distribution of (C+A)-rich sequences enriched by $(CA)_N$ imperfect repeats is highly skewed towards telomeres, and minisatellites can usually be found in the vicinity as well (Giraudeau *et al.*, 1999). Despite the listed properties associated with (G+T)-rich regions, they seem significantly under-investigated and may yet reveal unknown important functional properties in the near future.

2.2.5. *Alternated R/Y MRI regions adopt Z-DNA conformation*

Left-handed anti-parallel Z-DNA double helix conformation has been first characterized in 1979 by Wang and co-authors for $(GC)_3$ repeats (Wang *et al.*, 1979). Detailed Z-DNA structure has been considered elsewhere (Rich and Zhang, 2003; Ho, 2009). This particular conformation is characterized by rotation of R bases that adopt *syn* form and stack over the deoxyribose ring, while Y bases do not adopt unfavorable *syn* form (Ho, 2009). Thus, Z-DNA, which is characterized by alternating pattern of *anti-syn* conformations, is formed by alternating R/Y sequences (Johnston, 1992). The latest version of the *Genomic MRI* package has a new feature allowing the detection of excesses and shortages of alternating bases including R/Y patterns. It reveals that in mammalian genomes there is more than 40 times the over-abundance of alternating R/Y stretching over 50–100 bp genomic segments, where RY plus YR comprise more than 80% of all dinucleotides. A considerable portion of these alternating R/Y patterns are represented by short $(GC)_n$, $(AC)_n$, $(AT)_n$, and $(TG)_n$ repeats that can alternate with each other and be accompanied by alternating R/Y bases without strong periodic sequence pattern. For example, here is a sequence of a 50 bp segment from the third intron of human heparanase-2 gene highly enriched with alternated R and Y bases: 5'AAATGGATGTGTGTATATATATGAAGTCGATACACACACATATA CACATA3'. We showed that such alternating R/Y sequences are plentiful throughout the mammalian genomes either inside introns or within intergenic regions.

In 1986 Ho and others developed a ZHUNT program for detection of genomic sequences with high propensity to form Z-DNA (Ho *et al.*, 1986). They found a high concentration of these sequences near the transcription start sites (Schroth *et al.*, 1992; Rich and Zhang, 2003). Most recently, human genomic Z-DNA segments have been detected experimentally using a Z-DNA binding protein domain as a probe (Li *et al.*, 2009). These authors found an abundance of Z-DNA hotspots located in centromeres of 13 human chromosomes. Z-DNA-forming sequences induce high levels of genetic instability in both mammalian and bacterial cells. These sequences could be causative factors for gene translocations found in leukemias and lymphomas (Wang *et al.*, 2006). The discovery of certain classes of proteins bound to Z-DNA with high affinity and specificity indicated a biological role of this structure. Yet, it is a common view that Z-DNA is an unstable conformation that is formed and disappears during particular physiological activities such as transcription (Rich and Zhang, 2003).

2.3. Weak periodicities and loose patterns

In addition to MRI patterns, there are several weak genomic periodicities and specific signals at the mid-range scale. Many of them are described in "The codes of life" (Barbieri, 2008). Wherein, Trifonov reviewed different codes that exist in the genomes at DNA, RNA, and protein levels. He emphasized a special property of genomic sequences to make superposition (overlapping) of the codes they carry. The overlapping is possibly due to degeneracy of the codes and might be useful for organism survivability (Peleg *et al.*, 2004). Here we consider three types of such patterns in eukaryotic genomes.

2.3.1. Chromatin periodicities

There exists a nonrandom positioning of nucleosomes along genomic DNA of eukaryotes (Salih *et al.*, 2007). Nucleosome binding preferences are achieved via sequence-dependent deformational anisotropy of DNA (Barbieri, 2008). On average, one nucleosome occupies 200 bp including 145 nucleotides that contact its core particle while the rest corresponds to linkers between nucleosomes. Due to this specificity in nucleosome positioning, Trifonov and co-authors described sequence features that are repeated with 200- and 400-base periodicities (Trifonov 1998; Cohanim et al., 2006).

2.3.2. Periodicities in protein-coding sequences

There are well-known short-range periodicities in coding sequences that exist due to nonsymmetry in the genetic code, nonrandom amino acid appearance and association of neighboring amino acids within protein sequences, and also regularities in codon bias and context-dependent codon bias (Fedorov et al., 2002). In addition, there exist longer periodicities in the coding sequences that correspond to modular organization of globular proteins. They extend over 20–30 codons and represent initial protein folding modules (Aharonovsky and Trifonov, 2005; Barbieri, 2008).

2.3.3. Transcription-associated mutational asymmetry in mammals

In 2003 Green et al. demonstrated that the transcribed strands of mammalian DNA have an excess of G+T over A+C due to the difference of particular mutation frequencies (Green et al., 2003). Specifically, the A → G transition occurs at a 28% higher rate than the complementary transition T → C on the transcribed strand in most human genes. This transcription-associated mutational bias exists for both the exonic and intronic parts of genes. Thus, if we look at the nucleotide frequencies in the combined sequences of all human introns (T = 30.7%; A = 28.0%; G = 21.1%; C = 20.2%) there is a 3.6% excess of G+T over A+C. (These calculations were obtained on our nonredundant set of 11 315 human genes containing 96 931 introns (Bechtel et al., 2008). For each intron we removed the first 10 and the last 30 bases). We detected the same preference of G+T over A+C in introns of other mammals, a smaller preference for sea urchin (2.2%), and the highest preference in Arabidopsis (11.8%). For the mouse-ear cress, the nucleotide bias in introns is mainly due to significant excess of T (39.6%) over A (28.2%). Such strong transcriptional asymmetry in the preference of G+T over A+C is typical for other plants. No difference for G+T versus A+C composition has been detected for fruit fly and worm introns.

2.4. A complex mosaic of MRI patterns and their fundamental importance

2.4.1. Intricate arrangement of genomic MRI patterns

Different MRI regions are not randomly arranged relative to each other (Bechtel, 2008). For example, Fig. 6 illustrates that (G+C)-rich regions

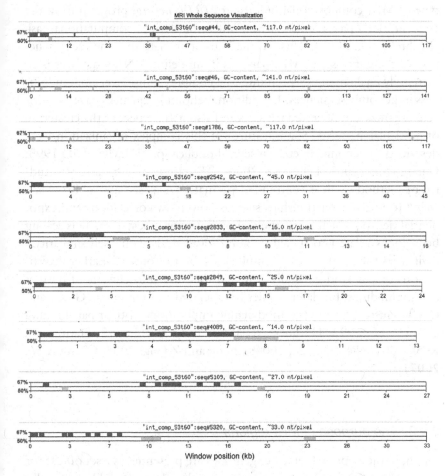

Fig. 6. Visualization of (G+C)-rich (dark gray, top row) and (A+T)-rich (light gray, bottom row) MRI features in human introns using a 400-nt base window size. The scale for each sequence is independent and is given in its subheading in nucleotides per pixel. The figure represents a fragment of Fig. 17 in Bechtel (2008).

tend to be associated in clusters. On the other hand, the distribution of (A+T)-rich regions is much more close to a random distribution with the exception that (A+T)-rich regions avoid very close proximity to each other (Bechtel, 2008). So far, investigators have examined only individual genomic patterns. The mutual arrangement of various genomic mid-range patterns has never been thoroughly investigated yet. Our preliminary results suggest that within mammalian genomes, there is a complex mosaic picture of MRI regions. Modeling sequences only with one particular type of MRI compositional bias using *MRI-generator* program from the *Genomic MRI* package has proven not to be a trivial computational task (Bechtel *et al.*, 2008). This has made us appreciate that the reconstruction of the entire set of MRI patterns in modeling DNA sequences is an extremely challenging mission due to a complex multi-layer nonrandomness in genomic sequences. In addition, genomic sequences have an intricate organization of nested patterns and also with respect to the clustering of particular patterns. Some features of this complex organization were described as genomic fractals in several publications (Havlin *et al.*, 1995; Cheng *et al.*, 2007; Pellionisz, 2008). This arrangement has been studied by methods such as "detrended fluctuation analysis" and a "Brownian walk" to uncover relationships such as power law correlations and exponential decays, which assess the scaling behavior of a system. This scaling behavior is related to fractal geometry and deals with "self-similarity," defined as the property of resembling a subset of oneself. Earlier investigations of this kind generally confined themselves to clusters of purines and pyrimidines, but later studies have shifted to examining (G+C) and (A+T) clusters for the thermodynamic implications of their pair-binding (Peng *et al.*, 1992; Havlin *et al.*, 1995; Peng *et al.*, 1995; Haring and Kypr, 2001; Nicolay *et al.*, 2004; Cheng and Zhang, 2005; Cheng *et al.*, 2007).

2.4.2. The purpose of MRI regions

Often, in the popular literature, genomes are presented as a set of texts or instructions. Such a representation implies that there should be an intelligent creature somewhere inside a cell interpreting these DNA texts. Thus, it is more appropriate to compare genomes with self-realization programs that autonomously fulfill their tasks and are able to respond to environment signals and conditions. Such programs must be extremely complicated for complex organisms, like humans, which are built from trillions of cells of hundreds of different kinds, yet sharing the same genomic sequence. There must be fundamental principles for construction and functioning of genomic programs. One of the most important principles is the Principle of Recursive Genome Function (PRGF) illuminated by Pellionisz (Pellionisz, 2008). The author considers the genome as an unsupervised operating system. The well-known examples of such a system are neural networks for which mathematical models describing their behavior have been developed. According to Pellionisz, "the recursive genome function is a process when at every step of development already-built proteins iteratively access sets of primary and ensuing auxiliary information packets of DNA to build constantly developing hierarchies of protein structures." In other words, there is a crucial flow of information from proteins back to the genomic DNA. According to Pellionisz, this principle converts a genome from a *closed* to an *open* physical system and resolves the paradox of genomic entropy posed by John Sanford (Sanford, 2005). This perspective elucidates the importance of MRI regions as specific sites for changing genomic information by proteins. Indeed, MRI regions are intricately associated with unusual DNA conformations, which in turn are binding sites for a number of proteins. These proteins could stabilize and/or initiate DNA conformation transformation and propagate the signal along neighboring DNA segments. For instance, Z-DNA binding proteins could initiate this transformation from right-handed B-DNA to the left-handed Z-form. This structural transition changes the DNA supercoiling for the regional DNA landscape and additionally creates specific B-Z-boundaries with flipped-over bases. Such transformation could modify, open, and/or hide, some information on the genomic DNA not only at the protein binding site but within neighboring regions.

3. Conclusions

Overall, within vast areas of previously thought "junk DNA," represented by introns and intergenic sequences, there exists an intricate mosaic of various MRI regions with extreme base compositions. Various genomic MRI regions are tightly associated with unusual DNA conformations and must be of crucial importance for proper functioning of multi-cellular eukaryotes. Understanding of genomic MRI functions is critical for the newly emerged field of personal genomics and also for drug discovery.

Acknowledgment

This material is based upon work supported by the National Science Foundation under Grant No. 0643542.

References

Abrescia NG, Gonzalez C, Gouyette C, Subirana JA. (2004) X-ray and NMR studies of the DNA oligomer d(ATATAT): Hoogsteen base pairing in duplex DNA. *Biochemistry* **43**: 4092–4100.

Aharonovsky E, Trifonov EN. (2005) Protein sequence modules. *J Biomol Struct Dyn* **23**: 237–242.

Bacolla A, Collins JR, Gold B *et al.* (2006) Long homopurine*homopyrimidine sequences are characteristic of genes expressed in brain and the pseudoautosomal region. *Nucleic Acids Res* **34**: 2663–2675.

Barbieri M. (2008) *The Codes of Life. The rules of Macroevolution. Biosemiontics.* Springer.

Bechtel JM. (2008) *Characterization of Genomic Mid-Range Inhomogenity.* pp. 97. Health Science Campus. University of Toledo, Toledo.

Bechtel JM, Wittenschlaeger T, Dwyer T *et al.* (2008) Genomic mid-range inhomogeneity correlates with an abundance of RNA secondary structures. *BMC Genomics* **9**: 284.

Benson G. (1999) Tandem repeats finder: A program to analyze DNA sequences. *Nucleic Acids Res* **27**: 573–580.

Berger I, Egli M, Rich A. (1996) Inter-strand C-H...O hydrogen bonds stabilizing four-stranded intercalated molecules: stereoelectronic effects of O4' in cytosine-rich DNA. *Proc Natl Acad Sci USA* **93**: 12116–12121.

Bernardi G. (2007) The neoselectionist theory of genome evolution. *Proc Natl Acad Sci USA* **104**: 8385–8390.

Blaszak RT, Potaman V, Sinden RR, Bissler JJ. (1999) DNA structural transitions within the PKD1 gene. *Nucleic Acids Res* **27**: 2610–2617.

Catasti P, Chen X, Mariappan SV, Bradbury EM, Gupta G. (1999) DNA repeats in the human genome. *Genetica* **106**: 15–36.

Chakraborty S, Sharma S, Maiti PK, Krishnan Y. (2009) The poly dA helix: A new structural motif for high performance DNA-based molecular switches. *Nucleic Acids Res* **37**: 2810–2817.

Cheng J, Tong ZS, Zhang LX. (2007) Scaling behavior of nucleotide cluster in DNA sequences. *J Zhejiang Univ Sci B* **8**: 359–364.

Cheng J, Zhang LX. (2005) Statistical properties of nucleotide clusters in DNA sequences. *J Zhejiang Univ Sci B* **6**: 408–412.

Choo KH. (1997) *The Centromere.* Oxford Univ Press, Oxford, UK.

Chou SH, Cheng JW, Fedoroff O, Reid BR. (1994) DNA sequence GCGAAT-GAGC containing the human centromere core sequence GAAT forms a self-complementary duplex with sheared G.A pairs in solution. *J Mol Biol* **241**: 467–479.

Chou SH, Chin KH, Wang AH. (2003) Unusual DNA duplex and hairpin motifs. *Nucleic Acids Res* **31**: 2461–2474.

Chou SH, Zhu L, Reid BR. (1994) The unusual structure of the human centromere (GGA)2 motif. Unpaired guanosine residues stacked between sheared G.A pairs. *J Mol Biol* **244**: 259–268.

Cohanim AB, Kashi Y, Trifonov EN. (2006) Three sequence rules for chromatin. *J Biomol Struct Dyn* **23**: 559–566.

Crawford DC, Acuna JM, Sherman SL. (2001) FMR1 and the fragile X syndrome: Human genome epidemiology review. *Genet Med* **3**: 359–371.

Cubero E, Abrescia NG, Subirana JA, Luque FJ, Orozco M. (2003) Theoretical study of a new DNA structure: The antiparallel Hoogsteen duplex. *J Am Chem Soc* **125**: 14603–14612.

Eickbush TH, Jamburuthugoda VK. (2008) The diversity of retrotransposons and the properties of their reverse transcriptases. *Virus Res* **134**: 221–234.

Fedorov A, Saxonov S, Gilbert W. (2002) Regularities of context-dependent codon bias in eukaryotic genes. *Nucleic Acids Res* **30**: 1192–1197.

Felsenfeld G, Rich A. (1957) Studies on the formation of two- and three-stranded polyribonucleotides. *Biochim Biophys Acta* **26**: 457–468.

Fernando H, Sewitz S, Darot J, Tavare S, Huppert JL, Balasubramanian S. (2009) Genome-wide analysis of a G-quadruplex-specific single-chain antibody that regulates gene expression. *Nucleic Acids Res* **37**: 6716–6722.

Gardiner-Garden M, Frommer M. (1987) CpG islands in vertebrate genomes. *J Mol Biol* **196**: 261–282.

Gibbs RA, Weinstock GM, Metzker ML *et al.* (2004) Genome sequence of the Brown Norway rat yields insights into mammalian evolution. *Nature* **428**: 493–521.

Giraudeau F, Petit E, Avet-Loiseau H, Hauck Y, Vergnaud G, Amarger V. (1999) Finding new human minisatellite sequences in the vicinity of long CA-rich sequences. *Genome Res* **9**: 647–653.

Goni JR, Vaquerizas JM, Dopazo J, Orozco M. (2006) Exploring the reasons for the large density of triplex-forming oligonucleotide target sequences in the human regulatory regions. *BMC Genomics* **7**: 63.

Grady DL, Ratliff RL, Robinson DL, McCanlies EC, Meyne J, Moyzis RK. (1992) Highly conserved repetitive DNA sequences are present at human centromeres. *Proc Natl Acad Sci USA* **89**: 1695–1699.

Green P, Ewing B, Miller W, Thomas PJ, Green ED. (2003) Transcription-associated mutational asymmetry in mammalian evolution. *Nat Genet* **33**: 514–517.

Greenbaum JA, Pang B, Tullius TD. (2007) Construction of a genome-scale structural map at single-nucleotide resolution. *Genome Res* **17**: 947–953.

Greenbaum JA, Parker SC, Tullius TD. (2007) Detection of DNA structural motifs in functional genomic elements. *Genome Res* **17**: 940–946.

Grokhovsky SL, Il'icheva IA, Nechipurenko DY, Panchenko LA, Polozov RL, Nechipurenko YD. (2008) Heterogeneity of DNA local structure and dynamics: ultrasound studies. *Biofizika* **53**: 417–425.

Hackenberg M, Previti C, Luque-Escamilla PL, Carpena P, Martinez-Aroza J, Oliver JL. (2006) CpGcluster: A distance-based algorithm for CpG-island detection. *BMC Bioinformatics* **7**: 446.

Hamilton BJ, Wang XW, Collins J *et al.* (2008) Separate cis-trans pathways post-transcriptionally regulate murine CD154 (CD40 ligand) expression: A novel function for CA repeats in the 3'-untranslated region. *J Biol Chem* **283**: 25606–25616.

Haring D, Kypr J. (2001) Mosaic structure of the DNA molecules of the human chromosomes 21 and 22. *Mol Biol Rep* **28**: 9–17.

Havlin S, Buldyrev SV, Goldberger AL *et al.* (1995) Statistical and linguistic features of DNA sequences. *Fractals* **3**: 269–284.

Herzel H, Weiss O, Trifonov EN. (1999) 10–11 bp periodicities in complete genomes reflect protein structure and DNA folding. *Bioinformatics* **15**: 187–193.

Ho PS. (2009) Methods to study nucleic acid structure. *Methods* **47**: 141.

Ho PS, Ellison MJ, Quigley GJ, Rich A. (1986) A computer aided thermodynamic approach for predicting the formation of Z-DNA in naturally occurring sequences. *EMBO J* **5**: 2737–2744.

Hui J, Hung LH, Heiner M *et al.* (2005) Intronic CA-repeat and CA-rich elements: A new class of regulators of mammalian alternative splicing. *EMBO J* **24**: 1988–1998.

Huppert JL. (2008) Four-stranded nucleic acids: Structure, function and targeting of G-quadruplexes. *Chem Soc Rev* **37**: 1375–1384.

Illingworth RS, Bird AP. (2009) CpG islands — 'a rough guide'. *FEBS Lett* **583**: 1713–1720.

Ioshikhes I, Trifonov EN, Zhang MQ. (1999) Periodical distribution of transcription factor sites in promoter regions and connection with chromatin structure. *Proc Natl Acad Sci USA* **96**: 2891–2895.

Jain A, Wang G, Vasquez KM. (2008) DNA triple helices: Biological consequences and therapeutic potential. *Biochimie* **90**: 1117–1130.

Jaishree TN, Wang AH. (1994) Human chromosomal centromere (AATGG)n sequence forms stable structures with unusual base pairs. *FEBS Lett* **347**: 99–103.

Johnston BH. (1992) Generation and detection of Z-DNA. *Methods Enzymol* **211**: 127–158.

Jurka J, Kapitonov VV, Kohany O, Jurka MV. (2007) Repetitive sequences in complex genomes: Structure and evolution. *Annu Rev Genomics Hum Genet* **8**: 241–259.

Katahira M, Fukuda H, Kawasumi H, Sugimura T, Nakagama H, Nagao M. (1999) Intramolecular quadruplex formation of the G-rich strand of the mouse hypervariable minisatellite Pc-1. *Biochem Biophys Res Commun* **264**: 327–333.

Kowalski D, Natale DA, Eddy MJ. (1988) Stable DNA unwinding, not "breathing," accounts for single-strand-specific nuclease hypersensitivity of specific (A+T)-rich sequences. *Proc Natl Acad Sci USA* **85**: 9464–9468.

Kypr J, Kejnovska I, Renciuk D, Vorlickova M. (2009) Circular dichroism and conformational polymorphism of DNA. *Nucleic Acids Res* **37**: 1713–1725.

Kypr J, Kejnovska I, Vorlickova M. (2007) Conformations of DNA strands containing GAGT, GACA, or GAGC tetranucleotide repeats. *Biopolymers* **87**: 218–224.

Lee C, Wevrick R, Fisher RB, Ferguson-Smith MA, Lin CC. (1997) Human centromeric DNAs. *Hum Genet* 100: 291–304.

Li H, Xiao J, Li J, Lu L, Feng S, Droge P. (2009) Human genomic Z-DNA segments probed by the Z alpha domain of ADAR1. *Nucleic Acids Res* 37: 2737–2746.

Liebich I, Bode J, Reuter I, Wingender E. (2002) Evaluation of sequence motifs found in scaffold/matrix-attached regions (S/MARs). *Nucleic Acids Res* 30: 3433–3442.

Liu R, Liu H, Chen X, Kirby M, Brown PO, Zhao K. (2001) Regulation of CSF1 promoter by the SWI/SNF-like BAF complex. *Cell* 106: 309–318.

Lupski JR. (1998) Genomic disorders: Structural features of the genome can lead to DNA rearrangements and human disease traits. *Trends Genet* 14: 417–422.

Malkov VA, Voloshin ON, Veselkov AG *et al.* (1993) Protonated pyrimidine-purine-purine triplex. *Nucleic Acids Res* 21: 105–111.

Mandel JL. (1993) Questions of expansion. *Nat Genet* 4: 8–9.

Matsugami A, Okuizumi T, Uesugi S, Katahira M. (2003) Intramolecular higher order packing of parallel quadruplexes comprising a G:G:G:G tetrad and a G(:A):G(:A):G(:A):G heptad of GGA triplet repeat DNA. *J Biol Chem* 278: 28147–28153.

Mirkin SM. (2008) Discovery of alternative DNA structures: A heroic decade (1979–1989). *Front Biosci* 13: 1064–1071.

Moss GP. Nomenclature for Incompletely Specified Bases in Nucleic Acid Sequences.

Nakagama H, Higuchi K, Tanaka E *et al.* (2006) Molecular mechanisms for maintenance of G-rich short tandem repeats capable of adopting G4 DNA structures. *Mutat Res* 598: 120–131.

Needle S. (2009) The structures of quadruplex nucleic acids and their drug complexes. *Curr Opin Struct Biol* 19: 239–250.

Nicolay S, Argoul F, Touchon M, d'Aubenton-Carafa Y, Thermes C, Arneodo A. (2004) Low frequency rhythms in human DNA sequences: A key to the organization of gene location and orientation? *Phys Rev Lett* 93: 108101.

Nonin-Lecomte S, Leroy JL. (2001) Structure of a C-rich strand fragment of the human centromeric satellite III: A pH-dependent intercalation topology. *J Mol Biol* 309: 491–506.

Oganesian L, Bryan TM. (2007) Physiological relevance of telomeric G-quadruplex formation: A potential drug target. *Bioessays* 29: 155–165.

Peleg O, Kirzhner V, Trifonov E, Bolshoy A. (2004) Overlapping messages and survivability. *J Mol Evol* 59: 520–527.

Pellionisz AJ. (2008) The principle of recursive genome function. *Cerebellum* 7: 348–359.

Peng CK, Buldyrev SV, Goldberger AL *et al.* (1995) Statistical properties of DNA sequences. *Physica A* 221: 180–192.

Peng CK, Buldyrev SV, Goldberger AL *et al.* (1992) Long-range correlations in nucleotide sequences. *Nature* 356: 168–170.

Prakash A, Shepard SS, Mileyeva-Biebesheimer O *et al.* (2009) Evolution of genomic sequence inhomogeneity at mid-range scales. *BMC Genomics* 10: 513.

Prokhortchouk E, Defossez PA. (2008) The cell biology of DNA methylation in mammals. *Biochim Biophys Acta* 1783: 2167–2173.

Radhakrishnan I, Patel DJ. (1993) Solution structure of a purine.purine.pyrimidine DNA triplex containing G.GC and T.AT triples. *Structure* 1: 135–152.

Rich A, Zhang S. (2003) Timeline: Z-DNA: the long road to biological function. *Nat Rev Genet* 4: 566–572.

Richard GF, Kerrest A, Dujon B. (2008) Comparative genomics and molecular dynamics of DNA repeats in eukaryotes. *Microbiol Mol Biol Rev* 72: 686–727.

Rigoutsos I, Huynh T, Miranda K, Tsirigos A, McHardy A, Platt D. (2006) Short blocks from the noncoding parts of the human genome have instances within nearly all known genes and relate to biological processes. *Proc Natl Acad Sci USA* 103: 6605–6610.

Salih F, Salih B, Trifonov EN. (2007) Sequence-directed mapping of nucleosome positions. *J Biomol Struct Dyn* 24: 489–493.

Sanford JC. (2005) Genetic Entropy & the Mystery of the Genome. Elim Publishing.

Schroth GP, Chou PJ, Ho PS. (1992) Mapping Z-DNA in the human genome. Computer-aided mapping reveals a nonrandom distribution of potential Z-DNA-forming sequences in human genes. *J Biol Chem* 267: 11846–11855.

Shogan B, Kruse L, Mulamba GB, Hu A, Coen DM. (2006) Virucidal activity of a GT-rich oligonucleotide against herpes simplex virus mediated by glycoprotein B. *J Virol* 80: 4740–4747.

Smit AFA, Hubley R, Green P. (1996–2008) RepeatMasker Open-3.1.8 <http://www.repeatmasker.org>.

Stefl R, Trantirek L, Vorlickova M, Koca J, Sklenar V, Kypr J. (2001) A-like guanine-guanine stacking in the aqueous DNA duplex of d(GGGGCCCC). *J Mol Biol* 307: 513–524.

Suzuki MM, Bird A. (2008) DNA methylation landscapes: provocative insights from epigenomics. *Nat Rev Genet* 9: 465–476.

Takai D, Jones PA. (2003) The CpG island searcher: a new WWW resource. *In Silico Biol* 3: 235–240.

Trifonov EN. (1998) 3-, 10.5-, 200- and 400-base periodicities in genome sequences. *Physica a-Statistical Mechanics and Its Applications* 249: 511–516.

Trifonov EN. (1991) DNA in profile. *Trends Biochem Sci* 16: 467–470.

Tsirigos A, Rigoutsos I. (2008) Human and mouse introns are linked to the same processes and functions through each genome's most frequent non-conserved motifs. *Nucleic Acids Res* 36: 3484–3493.

Umek RM, Linskens MH, Kowalski D, Huberman JA. (1989) New beginnings in studies of eukaryotic DNA replication origins. *Biochim Biophys Acta* 1007: 1–14.

Van Raay TJ, Burn TC, Connors TD *et al.* (1996) A 2.5 kb polypyrimidine tract in the PKD1 gene contains at least 23 H-DNA-forming sequences. *Microb Comp Genomics* 1: 317–327.

Vargason JM, Eichman BF, Ho PS. (2000) The extended and eccentric E-DNA structure induced by cytosine methylation or bromination. *Nat Struct Biol* 7: 758–761.

Vogt N, Rousseau N, Leng M, Malfoy B. (1988) A study of the B-Z transition of the AC-rich region of the repeat unit of a satellite DNA from Cebus by means of chemical probes. *J Biol Chem* 263: 11826–11832.

Wang AH, Quigley GJ, Kolpak FJ *et al.* (1979) Molecular structure of a left-handed double helical DNA fragment at atomic resolution. *Nature* 282: 680–686.

Wang G, Christensen LA, Vasquez KM. (2006) Z-DNA-forming sequences generate large-scale deletions in mammalian cells. *Proc Natl Acad Sci USA* **103**: 2677–2682.

Wang G, Zhao J, Vasquez KM. (2009) Methods to determine DNA structural alterations and genetic instability. *Methods* **48**: 54–62.

Wang L, Sommer M, Rajamani J, Arvin AM. (2009) Regulation of the ORF61 promoter and ORF61 functions in varicella-zoster virus replication and pathogenesis. *J Virol* **83**: 7560–7572.

Wang Y, Patel DJ. (1993) Solution structure of a parallel-stranded G-quadruplex DNA. *J Mol Biol* **234**: 1171–1183.

Warne SE, deHaseth PL. (1993) Promoter recognition by *Escherichia coli* RNA polymerase. Effects of single base pair deletions and insertions in the spacer DNA separating the -10 and -35 regions are dependent on spacer DNA sequence. *Biochemistry* **32**: 6134–6140.

Waterston RH, Lindblad-Toh K, Birney E *et al.* (2002) Initial sequencing and comparative analysis of the mouse genome. *Nature* **420**: 520–562.

Wells RD. (2007) Non-B DNA conformations, mutagenesis and disease. *Trends Biochem Sci* **32**: 271–278.

Yagil G. (2004) The over-representation of binary DNA tracts in seven sequenced chromosomes. *BMC Genomics* **5**: 19.

Chapter 4

Motif Finding from Chips to ChIPs

Giulio Pavesi*

Motif finding aimed at the *de novo* discovery of putative over-represented transcription factor binding sites in nucleotide sequences has been, and still is, one of the most challenging and open problems in bioinformatics. This article aims to provide a survey of different approaches and methods, from the days when typical instances were sets of promoters from co-expressed genes obtained from microarray (the "chips") expression data, to the latest advances in the field, that permit more reliable identification of transcription factor target sequences through genome-wide experiments like Chromatin Immunoprecipitation (the "ChIPs").

1. Introduction

Motif finding in bioinformatics can be defined as the problem of finding short similar sequence elements shared by a set of nucleotide or protein sequences with a common biological function, in order to single out which parts of the sequences are more likely to be essential for the function itself. The identification of regulatory elements in nucleotide sequences, modulating the expression of genes, has been one of the most widely studied flavors of the problem, both for its biological significance and for its sheer difficulty (Pavesi *et al.*, 2004; Sandve and Drablos, 2006).

It is a well-known fact that researchers in biology and medicine have nowadays at their disposal enormous amounts of data and information, like the complete DNA sequences of human and a number of different organisms of interest, that in turn permit the large-scale annotation of genes and their products, the bricks of which life is built. On the other hand,

---

*Dipartimento di Scienze Biomolecolari e Biotecnologie, Università di Milano, Milano, Italy. giulio.pavesi@unimi.it

introduction of technologies like oligonucleotide microarrays (Churchill, 2002; Schulze Downward, 2001) has given the possibility of measuring the level of transcription of genes, that is, when and how much a given gene is active according to developmental stage, cell cycle, external stimuli, disease, and so on.

This first step of gene expression, the transcription of a DNA region into a complementary RNA sequence, is finely modulated and regulated by the activity of *transcription factors* (TFs), which are proteins that in turn are encoded by the genome. TFs bind DNA in the neighborhood of the transcription start site of genes (in the *promoter* region), or often in distal elements (*enhancers* or *silencers*) that are brought by the 3D arrangement of DNA close to the gene region, with the effect of initiating the transcription process, or, in some cases, of blocking it (Lemon and Tjian, 2000; Levine and Tjian, 2003). The transcription of a given gene is thus initiated only when the "right" combination of TFs are bound to the DNA at the "right" time in its neighborhood. To understand the complexity of this process suffices it to say that about 10% of the about 22 000 human genes have been estimated to possess this function, that is, binding DNA in order to regulate the transcription of a subset of the other genes, with an exponential number of possible TF combinations and interactions.

The actual DNA region bound by a TF (called *transcription factor binding site*, or TFBS) usually ranges in size from 8–10 to 16–20 nucleotides (small sequence elements of this size are also called *oligonucleotides*, or *oligos*) (Lemon and Tjian, 2000; Stormo, 2000). TFs bind the DNA in a sequence-specific fashion, that is, they recognize sequences that are similar but not identical, tolerating a certain degree of "approximation." On the other hand, changing binding affinity according to DNA sequences allows TFs to obtain a more fine-grained modulating effect on the level of transcription of the target genes.

The "binding preference" of a given TF can be summarized and modeled starting from an experimentally validated collection of its sites, as shown in Fig. 1. In the simplest form, we can take, position by position, the most frequent nucleotide, and build a *consensus* of the sites. All oligos that differ from the consensus up to a maximum number of nucleotide substitutions can be considered valid instances of binding sites for the same TF. Clearly, this is an over-simplification that does not take into account the different level of variability at different position of the sites. A more involved method is to employ "degenerate consensuses," that can be formalized, for example, by using regular expressions. Positions where there does not seem to be any preference for any nucleotide are left "undefined," and any nucleotide is allowed there; positions where two or three nucleotides can be found with approximately the same frequency are left ambiguous, and any of the ambiguous nucleotides are considered a valid match; a single nucleotide is used only for the most conserved positions, which require an exact single-nucleotide match. Finally, the most flexible and widely used way of building descriptors for TF binding is to align the available sites, and to build an (ungapped) alignment profile with the frequency with which each nucleotide appears at each position in the sites. Thus, any candidate oligo can be compared to the profile, and the corresponding nucleotide frequencies can be used to assess how well it fits the descriptor (rather than a yes/no decision like with consensuses) (Stormo, 2000). A typical example is shown in Fig. 1. By comparing the different sites, we can notice that they differ in a few positions (mismatches) — as nearly always the situation with TFBSs. And, we can notice how some positions are strongly conserved, i.e. the TF does not seem to tolerate substitutions in those places, while in others any nucleotide seems to do.

Ever since researchers had at their disposal both genomic sequences and measurements of the level of transcription of genes through technologies like oligonucleotide microarray "chips," the rationale has been straightforward: TFs are responsible for activating or blocking the transcription of genes; genes co-expressed, that is, with similar expression patterns should be regulated to some extent by the same TFs; by investigating genomic regions taken from co-expressed genes (for example, their promoters) one should be able to detect the presence of short "over-represented" sequence elements, corresponding to binding sites for the common regulators (Brazma and Vilo, 2000; Cordero *et al.*, 2007).

On the other hand, while traditionally promoters from clusters of co-expressed genes have been the most typical input to algorithms for finding over-represented sequence motifs, more recently the introduction of technologies like Chromatin Immunoprecipitation (ChIP; Collas and Dahl, 2008), coupled with tiling arrays (ChIP on Chip; Pillai and Chellappan, 2009) or next-generation sequencing (ChIP-Seq; Mardis, 2007; Park, 2009), has permitted the direct genome-wide identification of regions bound *in vivo* by a given TF. As we will discuss in the following, these experiments are also a perfect case study for motif finding, and indeed have led to the introduction of novel methods inspired by them.

```
CTTGGTGACGTG
GTGAGTGACGTC
CGGGTTGACGCA
CCTACTTACGTA
TATGGTGACGTC
TCGGATGACGAT
TAGGATGACGTC      A  [ 0   3   0   2   5   0   0  16   0   0   1   5 ]
CCTGGTGACGCC      C  [ 7   5   3   3   1   0   0   0  16   0   5   6 ]
CGCGGTGACGTA      G  [ 5   4   6  11   7   0  15   0   0  16   0   3 ]
GCCGTTGACGCC      T  [ 4   4   7   0   3  16   1   0   0   0  10   2 ]
CGCGATGACGCA
CCTGTTGACGTG
TTGCATGACGTC
GTTGGTGACGTG
GAGGATGACGTT
GGTCGTGACGTA
------------
CTGGGTGACGTC  (Consensus)
------------
CNKGGTGACGTM  (Degenerate Consensus)
```

Fig. 1. Describing a "motif" representing the binding specificity of a transcription factor. Given a set of oligos known to be bound by the same TF, we can represent the motif they form by a "consensus" (bottom left) with the most frequent nucleotide in each position; a "degenerate" consensus, which accounts for ambiguous positions where there is no nucleotide clearly preferred (N = any nucleotide; K = G or T; M = A or C, according to IUPAC codes (Nomenclature Committee, 1986)); an alignment profile (right) which is converted into a nucleotide frequency matrix by dividing each column by the number of sites used.

Regardless of the source experiment, the problem can be informally defined as follows: given a set of DNA sequences, typically a few hundred base pairs (bps) long, find a set of oligos (10–16 bps long) appearing in all or most of the sequences (thus allowing for experimental errors and the presence of false positives in the set) similar to one another enough to be likely to be instances of sites recognized by the same TF. And, clearly, the same set of similar oligos should not appear with the same frequency and/or the same degree of similarity in a set of sequences selected at random (thus very unlikely to share any common regulator) or built at random with some model generating of "biologically feasible" DNA sequences. The similar and over-represented oligos collectively build a *motif* recurring in the input sequences.

The problem thus defined has been widely studied throughout all the history of bioinformatics, and all the methods introduced so far for the problem mainly differ in two points:

(1) in how similar oligos forming a candidate motif are chosen, and the motif they form is described and evaluated;

(2) in the "background random model" used to assess the statistical significance (over-representation) of the motifs.

The first choice is in how to model a solution, that is, summarizing a set of short sequence elements. As we have shown, there are two main strategies in describing the binding specificity of a TF: consensuses or profile matrices. Likewise, the most widely known motif discovery algorithms can be roughly split into "consensus-driven" and "profile-driven" methods (Pavesi *et al.*, 2004). In fact, we should keep in mind that, given k input sequences DNA of length n, and a motif size m, by assuming that a motif instance should appear in each sequence we have $(n - m + 1)^k$ candidate solutions that can be built by combining all the oligos of length m (m-mers) in all the possible ways, a number exponential in the number of input sequences that regardless of scoring function used to evaluate the solutions leads to a NP-hard problem. Leaving aside the design of performance-guaranteed approximation algorithms (Akutsu *et al.*, 2000), which produce solutions too much far from the optimal one to be biologically meaningful, different heuristics can be applied to the problem. Thus, the choice of how to model the motifs has straightforward implications in the heuristics that can be applied to explore the solution space.

2. Profile-Based Methods — The Basics

As stated before, profiles provide a more fine-grained and flexible description of the binding specificity of a TF. Unsurprisingly then, until the last few years nearly all the methods introduced for the problem have been profile-based. The basic idea is to select some oligos from the input sequences, align them, and score the resulting profile according to its conservation with a suitable measure of significance. Since exhaustive enumeration of all the possible profiles is computationally infeasible, methods of this kind have to rely on some heuristic to explore the solution space. Nearly all the "traditional" optimization techniques (i.e. greedy, local search, stochastic search, genetic algorithms, and so on) have been tried over the years. For sake of space, here we confine our description to those methods that have achieved greater success and wider usage in the biological community.

Despite its name, "Consensus" is an alignment-based method that employs a greedy heuristic (Hertz *et al.*, 1990; Hertz and Stormo, 1999).

Given as input a set of sequences $S_1 \ldots S_k$, the basic version of the algorithm requires as input the length m of the motif to be found, and assumes that it occurs once in each sequence. The steps performed by the algorithm can be summarized as follows: all the m-mers of S_1 are aligned to the m-mers of S_2, and each alignment produces a profile. All the profiles are scored according to their conservation, and the highest scoring ones are saved. Each oligo of length m of sequence S_3 is aligned with the matrices saved at the previous step, generating a new set of three-sequence profiles; each one is scored as before, and again only the highest scoring ones are saved. This step is iterated for each sequence of the set; the final profiles, output by the program, will contain one oligo for each input sequence. The algorithm is greedy, that is, at each step saves only the best partial alignments, and employs them to build new profiles. Further improvements to the algorithm introduced the possibility of finding motifs that do not occur or appear more than once in each sequence, and avoid explicitly requiring the length parameter from the user. Also, the calculation of a p-value for an alignment is introduced. The p-value gives an estimate of the probability of finding a profile with the same score by chance, which is especially useful in comparing alignments with different lengths and different numbers of sites.

Another way of looking at the problem of finding the best alignment profile is to assess whether a given oligo fits better the alignment profile or a "background" model against which the motif should stand out. Given a profile, the MEME (Multiple Expectation Maximisation for Motif Elicitation) algorithm (Bailey and Elkan, 1994; 1995) evaluates the likelihood of each oligo of given length to fit the profile with respect to the rest of the sequences, while the rest of the sequences should fit the background better than the profile. According to this principle, a likelihood normalized value is computed for each m-mer of each input sequence. This is the E (Expectation) step. Then, the algorithm builds a new profile by aligning all the sequence regions of length m, but weighting each one with the corresponding likelihood value. This is the M (Maximisation) step. At the beginning the algorithm builds a profile from each m-mer in the input sequences, then it performs on each one a single E and a single M step. The highest scoring profile obtained in this way is further optimized with more EM steps, until no further increase on the score is obtained. Finally, the profile is reported, and its oligos are removed from the input sequences. Then, the algorithm is restarted, until a given number of profiles that can be specified as input has been generated.

One of the most successful heuristics in profile-based motif finding has been the Gibbs sampling strategy, first introduced for motif discovery in protein sequences by Lawrence *et al.* (1993) and Neuwald *et al.* (1995) but nevertheless perfectly suitable also for nucleotide sequences. The best indicator of its success is the number of times it has been used in the algorithmic part of different methods, which varied the statistical measures used to generate and evaluate the results. The main motivation was to improve a EM local search strategy similar to the one employed by MEME (Lawrence and Reilly, 1990), avoiding possible premature convergence to local maxima, typical of local search heuristics. The basic idea, assuming again that one site appears in each input sequence, is the following:

(1) An m-mer is chosen at random in each of the k input sequences (at the beginning, with uniform probability).
(2) One of the k sequences is chosen at random: let S be this sequence.
(3) A profile is built with the oligos that had been taken from the other $k - 1$ sequences.
(4) A likelihood value is computed for each oligo in S, representing how well it fits the model induced by the profile with respect to some background distribution.
(5) A new probability value, proportional to the likelihood, is assigned to each oligo of S.
(6) Go to the first step, applying it to sequence S only: now the probability with which the m-mers of sequence S can be picked are those computed at the previous step, and the oligos that fit better in the alignment described by the profile are more likely to be chosen, and to replace the original oligo chosen with uniform probability.

These steps are iterated a number of times, or until convergence is reached. This variant of the algorithm is also known as the *site sampler*. At the beginning all the oligos have the same probability of being chosen; in successive iterations, those that better fit the profile are more and more likely (but not guaranteed) to be selected. The main difference with MEME is in the first step: while local search selects oligos deterministically according to how well they fit the current solution, the Gibbs sampler chooses how to modify the current solution in a stochastic way. The algorithm is thus less

likely to get stuck in local optima; on the other hand, given its probabilistic
nature, it has often to be run several times.

Further improvements were introduced successively (Neuwald *et al.*,
1995), allowing multiple occurrences (or no occurrence) of a motif within
the same sequence (algorithm known as *motif sampler*). Modifications of
the basic Gibbs sampling technique especially devised for DNA sequences
are described in Hughes *et al.* (2000) and Workman and Stormo (2000).
AlignACE (Hughes *et al.*, 2000) is a fine-tuned Gibbs sampling algorithm for DNA regulatory sequences, including for example a different
sampling technique that also considers similarity in the position in the
input sequences of each of the oligos of a motif. In ANN-Spec (Workman
and Stormo, 2000), Gibbs sampling is combined with an artificial neural
network that replaces the frequency matrix. Instead of aligning the oligos, the algorithm trains a neural network in order to recognize the oligos
selected (the putative TFBSs) against the rest of the sequences.

3. Profile-Based Methods — Modeling the Background

So far, we have described different optimization strategies leaving aside the
discussion of the significance function to be optimized. Quite intuitively,
the latter should simultaneously take into account both how much each
column of the profile is conserved and how the nucleotide frequencies in
the columns of the profile differ from a "background" distribution that
should be obtained from aligning random sequences. If we assume that
nucleotides in genomic sequences are independent, that is, the probability
of finding a nucleotide in any position is not influenced by its neighbors,
the overall conservation (similarity among the oligos) of the motif and its
distance from the "background noise" can be measured by computing the
information content (IC) or *relative entropy* of the profile:

$$IC = \sum_{i=1}^{4} \sum_{j=1}^{m} m_{i,j} \log \frac{m_{i,j}}{b_i},$$

where $m_{i,j}$ is entry in row i and column j of the profile and b_i is the
expected frequency of nucleotide i in the input sequences (which in turn
can be estimated by using the genomic sequence of the organism studied or by the input sequences themselves). Clearly, for each column j we

have that $\sum_{i=1}^{4} m_{i,j} = 1$ and $\sum_{i=1}^{4} b_i = 1$. It can be clearly seen how this measure accounts for how much each column is conserved, and how much the nucleotide frequencies obtained in the profile differ from what would have been obtained by aligning oligos chosen at random. Notice that in case of uniform background frequencies, this measure equals Shannon's entropy, where the source probabilities equal the nucleotide frequencies in the profile. Information content was the measure used in the first versions of Consensus, MEME and the Gibbs Sampler. It is suitable for comparing alignments built by using the same number of sequences, and in order to compare alignments containing a different number of oligos, the information content of each profile can be multiplied by the number of oligos aligned, yielding the maximum a posteriori (MAP, or log-likelihood ratio) score.

The main drawback of this type of model is the independence assumption: in other words, the probability associated with each nucleotide in the background is not influenced by its neighbors in the sequence, an assumption that can be easily proven to be too restrictive. It is thus hardly a surprise that in the "second generation" of profile-based methods, better results were obtained not by introducing a novel optimization heuristic, but rather by focusing on the statistical measure of significance and especially the background model.

After all, the better the representation of the background is, the more likely a method is to detect something that significantly differs from the background itself, and the strategy used plays a less essential role. Thus, an improvement, introduced in different tools, has been to model the background with a higher-order Markov model (Thijs *et al.*, 2001). Intuitively, when a j-th order Markov model is employed, the probability of finding a nucleotide in a given position of a sequence depends on the j nucleotides preceding it in the sequence itself. Indeed, recent research has also focused on which model order seems to be more suitable for the problem. The model parameters can be estimated from the analysis of a number of regulatory regions of different species (for example, by taking all the promoters of all the genes annotated in a given species), leading to organism-specific probability distributions and expected oligo frequencies (Marchal *et al.*, 2003). In turn, the IC and MAP scores can be augmented by terms indicating not only how much the profile itself is conserved, but also how "surprising" it is to find the oligos composing it in the sequences analyzed according to the background model employed (Narasimhan *et al.*, 2003).

For example, the performance of MEME can be significantly improved by the introduction of an higher-order background (Bailey *et al.*, 2006). Bioprospector (Liu *et al.*, 2001) is a Gibbs sampler where the background is described with a third-order Markov model based on the genome-wide analysis of different organisms, and also Thijs *et al.* (2002) and Aerts *et al.* (2003) present similar algorithms where higher-order organism-specific modeling of the background is employed.

GLAM (Frith *et al.*, 2004) is another Gibbs sampler especially tailored to TFBSs, where the sampling procedure as well as the IC score have been modified in order to compare profiles of different size, sparing the user the visual inspection and comparison of the results obtained on different lengths. The optimal motif length is computed with a simulated annealing strategy.

Also the authors of the original Gibbs sampler have developed versions of their algorithm designed and adapted for TFBSs (Thompson *et al.*, 2003), with an improved method for the modeling of the sequence background. The idea is to use position-specific frequencies: in other words, if an oligo is located 100 bps upstream of the transcription start site of a gene, its expected frequency is estimated by analyzing the oligos that appear at approximately the same distance from the transcription start site with a Bayesian segmentation algorithm.

NestedMICA (Down and Hubbard, 2005) introduces mosaic background modeling. The idea is to use four different higher-order background models according to the overall nucleotide composition of the input sequences, and in particular to the content of C and G nucleotides (corresponding to the presence or absence of CpG islands in promoters). The profile optimization strategy adopted in this algorithm is also novel, based on a sequential Monte Carlo Expectation Maximization approach.

Indeed, work has been done also on the optimization strategies. It should be noticed that in nearly all the algorithms, we just mentioned the optimization steps are performed only by adding or replacing oligos in the current solution according to their respective similarity or their fitness in the profile, but the IC, MAP or similar scores are optimized only *a posteriori* by comparing different candidate solutions. A straightforward approach could be then to optimize directly the scoring function, as in Fogel *et al.* (2004) where evolutionary computation is employed. A recent work (Defrance and Helden, 2009) applies the Gibbs sampling strategy directly to the IC score of the profile to be optimized, reporting performance improvement with respect to traditional *a posteriori* methods.

4. Consensus-Based Methods — The Basics

As described earlier, a set of oligos can be summarized by using a consensus, and all the oligos differing from the consensus up to a maximum number of mismatches can be considered *a priori* a valid instance of the motif. Hence, the problem can be formalized as follows: for each of the 4^m DNA strings of TFBS-like length m (8–16 bps), collect from the input sequences all its approximate occurrences with up to e mismatches. In other words, the problem becomes exhaustive approximate pattern matching, allowing typically from two to four substitutions, according to the motif length. Introduced in the early days of bioinformatics (Galas *et al.*, 1985; Sadler *et al.*, 1983; Waterman *et al.*, 1984), this approach had been abandoned because it was considered too time-consuming, since it required enumerating an exponential (in the solution length) number of candidate solutions. The application of indexing structures to the input sequence set showed later on its feasibility, reducing its complexity to be exponential in the number of mismatches allowed (Pavesi *et al.*, 2001; Marsan and Sagot, 2000).

The simplest solution is anyway to consider only exact oligos, that is, allow no substitutions in instances of the same motif. The problem thus becomes much simpler and its complexity is just linear in the length of the input. Given its computational efficiency, this strategy can be employed in genome-wide analyses of over-represented oligos, as for example in (Caselle *et al.*, 2002; Cora *et al.*, 2004; van Helden *et al.*, 1998), among many others. Similar over-represented sequences can anyway be clustered in a post-processing stage, and considered different forms of binding sites for the same TF.

The search space can be also trimmed down by using "degenerate consensuses" to model the solutions, and tolerate mismatches only in the degenerate or ambiguous positions, with a significant improvement in the time needed for matching (Shinozaki *et al.*, 2003; Sinha and Tompa, 2003). The YMF algorithm (Sinha and Tompa, 2003) computes the expected number of occurrences of each motif with a Markov model of 4th order, and evaluates its significance with a statistical z-score.

No restrictions on the position of mutations are imposed by the SMILE (Marsan and Sagot, 2000) and Weeder (Pavesi *et al.*, 2001; 2004) algorithms, where the exhaustive search for the exponential number of candidate consensuses is implemented with the preliminary indexing of the sequences with a suffix tree. While the structure underlying the algorithm is virtually the same, the two approaches differ in how the significance of the motifs found is evaluated. SMILE compares the number of occurrences of a given motif with its occurrences in a random set of sequences of the same size built with a Markov model whose parameters are estimated from the input. Alternatively, a negative set of sequences that should not contain any instance of the binding sites appearing in the positive set can be used: the highest-scoring motifs will be the ones that present the most significant variation between the number of occurrences in the input set and in the random or negative sets.

In Weeder, instead the observed number of occurrences of a motif is compared with expected frequencies derived from the oligo-frequency analysis of all the regulatory regions of the same organism of the input sequences. The significance score is the sum of a general term and a sequence-specific term, based respectively on how many sequences each motif appears in and how much conserved it is in each sequence. Motif parameters (length, number of substitutions) suitable for TFBS identification are automatically employed by the algorithm in different runs. A post-processing phase compares the top-scoring motifs of each run in order to detect which ones could be more "interesting." Finally, the best instances of each motif are selected from the sequences by using a profile built with the oligos selected by the consensus-based algorithm, in order to have a more fine-grained ranking of the oligos that belong to the motif, and the possibility of detecting oligos well fitting the motif but exceeding the predefined substitution thresholds used.

A further refinement of consensus-associated significance measures is presented in Marschall and Rahmann (2009), where given a Markov model of some order, a *p*-value is computed with a compound Poisson approximation for the null distribution of the number of motif occurrences both in terms of overall number of occurrences and of number of sequences containing a motif instance. The monotonicity properties of the compound Poisson approximation are exploited to avoid exhaustive enumeration of candidate consensuses.

5. Other Methods

Clearly, profiles and consensuses are not the only possible ways to model a set of oligos. While the representation is relevant when one has to employ it to predict candidate sites according to the model (and this is what is done in profile- or consensus-based methods for selecting oligos and build motifs from them), it is much less of an issue when the aim is detecting a set of oligos sharing some level of pairwise similarity. Starting from this observation, another straightforward way of modeling the problem is to employ a graph, where nodes correspond to the oligos of the input sequences and edges connect nodes corresponding to similar enough oligos. The problem can be thus recast in graph-theory terms, and motifs can be found for example by detecting cliques (Pevzner and Sze, 2000) or maximum density subgraphs (Fratkin *et al.*, 2006).

6. Does Motif Finding Work?

It is far from being straightforward to evaluate the merits and shortcomings of the different algorithms for motif finding. Building sequence sets on which the "answer" is already known is quite hard, since the experimental validation *in vivo* of a given candidate site is a slow and painstaking work, and much more so assembling a promoter set of feasible size (for motif finding algorithms). Algorithms are thus often tested on synthetic datasets, in which "simulated" binding sites are planted into "simulated" sequences (Pevzner and Sze, 2000; Buhler and Tompa, 2002; Hu, 2003; Sze *et al.*, 2002).

Some benchmark sequence sets derived from experimental data were then introduced over the last few years, like the one used in the comparative assessment presented in Tompa *et al.* (2005), Li and Tompa (2006), or in Sandve *et al.* (2007), which are still very often used to assess the performance of a novel method.

All in all, the overall picture that emerged was that motif finding algorithms could provide reliable results in simple organisms like bacteria or yeast, but that the situation in higher eukaryotes like human was much bleaker. Consensus-based methods showed somehow a slightly better performance (Tompa *et al.*, 2005; Sinha and Tompa, 2003), probably thanks to the possibility of performing an exhaustive search and thus of finding optimal solutions, or sub-optimal candidate solutions to be further refined with profile-based optimization methods.

Apart from the design of benchmark sets, the poor performance in motif finding in practice can be due to several reasons. First of all, the fact that the similarity shared by sites recognized by the same TF is often very subtle, and when just a few sequences are investigated the motif they form is not conserved enough to be discriminated against background noise, and vice versa random similarities are likely to be detected and reported as a motif because the similarity is "more significant." Second, we have to consider the fact that the complexity of the regulation of every level of gene expression seems to grow in parallel to the complexity of the organisms, and thus also in the regulation of transcription. In other words, the observation of a set of co-expressed genes in human, mouse, or Drosophila is most often than not the result of the combined activity of several different co-operating or competing factors, each one acting on a subset of genes. Hence, what would be "interesting" motifs are much less over-represented or not over-represented at all regardless of the model employed. Third, the analysis of the promoter region alone is often not enough, since transcription can be regulated by distal elements like enhancer or silencers: thus motifs are not found simply because they do not appear at all in the sequences investigated, but rather thousands or even millions of base pairs away from the genes they regulate, a sequence size clearly out of bounds for motif finding algorithms. Fourth, we should not forget the fact that DNA is, after all, a molecule, with a three-dimensional structure it assumes by wrapping around histones forming chromatin. Hence, not all the regions of DNA are accessible to TFs, but their binding depends first of all on chromatin structure and its modifications.

Other additional considerations are thus usually employed when performing a standard promoter analysis aimed at finding the common regulators of the genes and their sites in the sequences. Perhaps the most widely used technique is phylogenetic footprinting (Chiang et al., 2003; Sauer et al., 2006; Dermitzakis and Clark, 2002), that is, to compare a given sequence with its orthologous counterparts in evolutionary close enough species. In fact, not only a strikingly high level of sequence conservation can be observed by comparing the protein coding sequence of orthologous genes from, for example, human and mouse, but also in non-coding regions likely to possess some regulatory function (Bejerano et al., 2004). Identifying significantly conserved non-coding sequences in different genomes has thus become the method of choice for example for predicting likely distal regulatory elements like enhancers or silencers, but also for standard promoter analysis, by identifying conserved short sequence elements likely to be single TFBSs conserved by evolution. Quite naturally this type of analysis can be performed prior to motif finding, by masking out the less conserved parts of the promoters investigated, but also simultaneously, by designing algorithms aimed at finding motifs both over-represented in a sequence set, and at the same time significantly conserved with respect to homologous sequences in other species (see for example Siddharthan *et al.* (2005) and Sinha *et al.* (2004), which are enhancements of the Gibbs sampler and MEME to this case). The drawback is clearly that we cannot expect, for example, to have every single functional site in human to be conserved in other species (Odom *et al.*, 2007).

Improvements in motif finding have also been reported when for example nucleosome occupancy of sequences (that is, selecting for analysis only those parts of the sequences that are free to be bound by TFs) has been added to the input (Narlikar *et al.*, 2007). In any case, a very elegant way to incorporate any kind of additional information in any motif finding algorithm is to modify the *a priori* probabilities, as in Tang *et al.* (2008), where a sampling method (parallel tempering) similar to NestedMICA is employed with better convergence properties than standard Gibbs sampling. While in the general formalization of the problem, all the oligos in the input sequences have the same probability of being part of a conserved motif, these probabilities can be modified according to any given criterion, for example, conservation in orthologous sequences giving conserved oligos higher *a priori* probabilities, and consequently higher probability of being selected during the optimization iterations.

7. Chips vs ChIPs

In the last few years, novel experimental methodologies have been introduced, opening to researchers in the field novel avenues of unprecedented power. Modern laboratory techniques, in fact, allow for the large-scale identification of TF-DNA binding sites on the genome, with experiments that were simply impossible to perform just a few years ago. A striking example is *Chromatin Immunoprecipitation* (ChIP) (Collas and Dahl, 2008), that allows for the extraction from the cell nucleus of a specific protein–DNA chromatin complex, in our case of a given TF bound *in vivo* to the DNA. First, the TF is cross-linked, that is, fixed to the DNA. Then, a specific antibody that recognizes only the TF is employed, and the antibody, bound to the TF which in turn is bound to the DNA, permits the extraction and isolation of the chromatin complex. At this point, DNA is released from the TF by reverse-crosslinking — and researchers have at their disposal the DNA regions corresponding to the genomic locations of the sites that were bound *in vivo*, that is, inside living cells. The experiment is performed on thousands of cells at the same time, so to have a quantity of DNA suitable for further analysis, and to have in the sample a good coverage of all regions bound by the TF.

The next phase is quite logically the identification of the DNA regions themselves — and of their corresponding location in the genome, which in turn is made possible by the availability of the full genomic sequences. Also for this step, technology has witnessed dramatic improvements. From the identification of only pre-determined candidate sites through PCR, the introduction of "tiling arrays" has permitted the analysis of the DNA extracted on a whole-genome scale (ChIP on Chip; Pillai and Chellappan, 2009) by using probes designed to cover the sequence of a whole genome. The recent introduction of novel and efficient sequencing technologies collectively known as *next-generation sequencing* (Mardis, 2008) has permitted to move this type of experiment one step further, by providing at reasonable cost perhaps the simplest solution: in order to identify the DNA extracted by the cell, simply sequence the DNA itself (ChIP Sequencing, or ChIP-Seq (Mardis, 2007)) and compare it to the reference genomic sequence. Once the regions have been identified, the next logical step is the identification of the actual oligos bound by the TF investigated, which is much shorter than the region itself — and thus another perfect case study for motif finding.

Luckily enough, when instead of a set of promoters from co-expressed genes, the input is a set of sequences derived from a ChIP experiment, the problem seems to become easier, for different reasons:

- The size of the sequence sets: a genome-wide ChIP usually yields thousands of candidate regions, while a typical promoter analysis of co-expressed genes is performed on a few dozen sequences.
- The length of the sequences: sequences extracted from experiments like ChIP-Seq can be as short as 200–300 bp, as opposed to promoters which are usually defined as 500–1000 bp long.

• The frequency with which binding sites for the same TF appear: in a ChIP there should appear a very high percentage of the sequences examined, even more than once in a single sequence, while in a set of co-expressed genes there is no guarantee for this.

All in all what happens is that with ChIP we have a somewhat cleaner input sequence set (most of the sequences should contain an instance of the motif), and more importantly much more redundant, since in thousands of sequences we can expect to find several instances of binding sites very similar to one another. In gene promoter analysis, the input set is much less cleaner (there is no guarantee on how many sequences actually share the same motif), the sequence set is much smaller (and thus the different sites in the sequences can be very different from one another) and the sequences are longer. Thus, in ChIP we can expect to obtain a clearer separation of the signal (the motif) from the background "noise."

Indeed, the performance of traditional motif finding methods on ChIP sequence sets managed to redeem their bad reputation in several cases, by actually "discovering" the sites bound by the TF (see among many others: Krig *et al.*, 2007; Zeller *et al.*, 2006; Chen *et al.*, 2008; Loh *et al.*, 2006). On the other hand, algorithms devised ad hoc for large scale ChIP experiments like MDScan (Liu, *et al.*, 2002), Trawler (Ettwiller *et al.*, 2007) and Amadeus (Linhart *et al.*, 2008) have been introduced, with somewhat superior performance in terms of motifs correctly identified, but much more significantly in computational resources required over methods that were devised for smaller sequence sets and subtler motif instances. The general ideas underlying these three methods are somewhat similar. Initial candidate solutions are built by matching consensuses (MDScan) or degenerate consensuses on the input sequences — indexed with a suffix tree in Trawler to speed up the search. Exact or degenerate consensuses are in fact powerful enough to capture significant motifs given the much higher redundancy of motif instances in the sequences. Significance is then assessed with a third-order background model (MDScan), or more simply by comparing the match counts in the input to randomly selected background sequence sets, with z-scores (Trawler) or a hypergeometric test (Amadeus). Finally, similar motifs are merged, and motifs are modeled with a profile which is further optimized on the input sequences.

Research on this topic is also trying to take advantage of the additional features that can be associated to regions derived from ChIP. For example, since probe or sequence enrichment defining a bound region in ChIP is reported to be an indicator of the affinity of the TF for the region (Jothi *et al.*, 2008), higher priority or weight should be given to those regions that are more enriched in the experiment. This is something that can be trivially done by applying existing methods to only the "best" sequences, but this factor could be taken into account directly by the algorithms, as in MDScan, similarly to what has been done by correlating sequence motifs in promoters with gene expression (Bussemaker *et al.*, 2001; Roven and Bussemaker, 2003). The modeling of the binding specificity of the TF with a profile (resulting from motif discovery algorithms) should also give higher weight to high-affinity sites corresponding to the most enriched regions. Finally, in experiments like ChIP-Seq the sites bound by the TF are more likely to be located in the center of the region extracted (Jothi *et al.*, 2008): thus adding positional bias to sequence conservation in assessing the motifs should somewhat improve the results.

Apart from their sheer interest for the scientific community, genome-wide ChIP experiments, now quite common, provide also a source of great value for building feasible benchmark sequence sets for the testing of motif finding algorithms, like the "Harbison dataset" derived from the work on 203 DNA binding proteins in yeast presented in Harbison *et al.* (2004), or the "metazoan dataset" introduced in Linhart *et al.* (2008), built from several different genome-wide ChIP experiments.

8. Conclusions

Motif finding has been and still is one of the most challenging problems in bioinformatics. Without claiming to be exhaustive, in this article we provided a survey of different methods and approaches for the problem, applied to the discovery of over-represented candidate transcription factor binding sites in nucleotide sequences. We discussed only the most general definition of the problem, while there exist other different widely studied flavors of it, from phylogenetic footprinting to the detection of correlated or structured motifs to the discovery of clusters of motifs forming cis-regulatory modules (CRMs) like enhancers or silencers. From every point of view, however, we are very far from the common goal of having a complete and thorough annotation of sites for transcription factors, in virtually all the genomes of interest. Thus, research in this area is still very active, and we can imagine it will remain so in the next few years. We advise the interested reader to keep a close eye on the bioinformatic advances in this field, but most importantly to follow the evolution of experimental techniques: for example the scenario changed radically when ChIP was introduced, and with its coupling with next-generation sequencing technologies changed even more. After all, selecting a good problem to work on is at least as important as finding good solutions for it.

References

Aerts S, Thijs G, Coessens B, Staes M, Moreau Y, De Moor B. (2003) Toucan: Deciphering the cis-regulatory logic of coregulated genes. *Nucleic Acids Res* **31**: 1753–1764.

Akutsu T, Arimura H, Shimozono S. (2000), *RECOMB 2000*. ACM, Tokyo, pp. 1–12.

Bailey TL, Elkan C. (1994) Fitting a mixture model by expectation maximization to discover motifs in biopolymers. *Proc Int Conf Intell Syst Mol Biol* **2**: 28–36.

Bailey TL, Elkan C. (1995) The value of prior knowledge in discovering motifs with MEME. *Proc Int Conf Intell Syst Mol Biol* **3**: 21–29.

Bailey TL, Williams N, Misleh C, Li WW. (2006) MEME: Discovering and analyzing DNA and protein sequence motifs. *Nucleic Acids Res* **34**: W369–373.

Bejerano G, Pheasant M, Makunin I *et al.* (2004) Ultraconserved elements in the human genome. *Science* **304**: 1321–1325.

Brazma A, Vilo J. (2000) Gene expression data analysis. *FEBS Lett* **480**: 17–24.

Buhler J, Tompa M. (2002) Finding motifs using random projections. *J Comput Biol* **9**: 225–242.

Bussemaker HJ, Li H, Siggia ED. (2001) Regulatory element detection using correlation with expression. *Nat Genet* **27**: 167–171.

Caselle M, Di Cunto F, Provero P. (2002) Correlating overrepresented upstream motifs to gene expression: A computational approach to regulatory element discovery in eukaryotes. *BMC Bioinformatics* **3**: 7.

Chen X, Xu H, Yuan P *et al.* (2008) Integration of external signaling pathways with the core transcriptional network in embryonic stem cells. *Cell* **133**: 1106–1117.

Chiang DY, Moses AM, Kellis M, Lander ES, Eisen MB. (2003) Phylogenetically and spatially conserved word pairs associated with gene-expression changes in yeasts. *Genome Biol* **4**: R43.

Churchill GA. (2002) Fundamentals of experimental design for cDNA microarrays. *Nat Genet* **32** Suppl: 490–495.

Collas P, Dahl JA. (2008) Chop it, ChIP it, check it: The current status of chromatin immunoprecipitation. *Front Biosci* **13**: 929–943.

Cora D, Di Cunto F, Provero P, Silengo L, Caselle M. (2004) Computational identification of transcription factor binding sites by functional analysis of sets of genes sharing overrepresented upstream motifs. *BMC Bioinformatics* **5**: 57.

Cordero F, Botta M, Calogero RA. (2007) Microarray data analysis and mining approaches. *Brief Funct Genomic Proteomic* **6**: 265–281.

Defrance M, van Helden J. (2009) info-gibbs: A motif discovery algorithm that directly optimizes information content during sampling. *Bioinformatics* **25**: 2715–2722.

Dermitzakis ET, Clark AG. (2002) Evolution of transcription factor binding sites in Mammalian gene regulatory regions: Conservation and turnover. *Mol Biol Evol* **19**: 1114–1121.

Down TA, Hubbard TJ. (2005) NestedMICA: Sensitive inference of over-represented motifs in nucleic acid sequence. *Nucleic Acids Res* **33**: 1445–1453.

Ettwiller L, Paten B, Ramialison M, Birney E, Wittbrodt J, (2007) Trawler: *de novo* regulatory motif discovery pipeline for chromatin immunoprecipitation. *Nat Methods* **4**: 563–565.

Fogel GB, Weekes DG, Varga G *et al.* (2004) Discovery of sequence motifs related to coexpression of genes using evolutionary computation. *Nucleic Acids Res* **32**: 3826–3835.

Fratkin E, Naughton BT, Brutlag DL, Batzoglou S. (2006) MotifCut: Regulatory motifs finding with maximum density subgraphs. *Bioinformatics* **22**: e150–157.

Frith MC, Hansen U, Spouge JL, Weng Z. (2004) Finding functional sequence elements by multiple local alignment. *Nucleic Acids Res* **32**: 189–200.

Galas DJ, Eggert M, Waterman MS. (1985) Rigorous pattern-recognition methods for DNA sequences. Analysis of promoter sequences from *Escherichia coli. J Mol Biol* **186**: 117–128.

Harbison CT, Gordon DB, Lee TI *et al.* (2004) Transcriptional regulatory code of a eukaryotic genome. *Nature* **431**: 99–104.

Hertz GZ, Hartzell GW, Stormo GD. (1990) Identification of consensus patterns in unaligned DNA sequences known to be functionally related. *Comput Appl Biosci* **6**: 81–92.

Hertz GZ, Stormo GD. (1999) Identifying DNA and protein patterns with statistically significant alignments of multiple sequences. *Bioinformatics* **15**: 563–577.

Hu YJ. (2003) Finding subtle motifs with variable gaps in unaligned DNA sequences. *Comput Methods Programs Biomed* **70**: 11–20.

Hughes JD, Estep PW, Tavazoie S, Church GM. (2000) Computational identification of cis-regulatory elements associated with groups of functionally related genes in *Saccharomyces cerevisiae. J Mol Biol* **296**: 1205–1214.

Jothi R, Cuddapah S, Barski A, Cui K, Zhao, K. (2008) Genome-wide identification of *in vivo* protein-DNA binding sites from ChIP-Seq data. *Nucleic Acids Res* **36**: 5221–5231.

Krig SR, Jin VX, Bieda MC *et al.* (2007) Identification of genes directly regulated by the oncogene ZNF217 using chromatin immunoprecipitation (ChIP)-chip assays. *J Biol Chem* **282**: 9703–9712.

Lawrence CE, Altschul SF, Boguski MS, Liu JS, Neuwald AF, Wootton JC. (1993) Detecting subtle sequence signals: A Gibbs sampling strategy for multiple alignment. *Science* **262**: 208–214.

Lawrence CE, Reilly AA. (1990) An expectation maximization (EM) algorithm for the identification and characterization of common sites in unaligned biopolymer sequences. *Proteins* **7**: 41–51.

Lemon B, Tjian R. (2000) Orchestrated response: A symphony of transcription factors for gene control. *Genes Dev* **14**: 2551–2569.

Levine M, Tjian R. (2003) Transcription regulation and animal diversity. *Nature* **424**: 147–151.

Li N, Tompa M. (2006) Analysis of computational approaches for motif discovery. *Algorithms Mol Biol* **1**: 8.

Linhart C, Halperin Y, Shamir R. (2008) Transcription factor and microRNA motif discovery: The Amadeus platform and a compendium of metazoan target sets. *Genome Res* **18**: 1180–1189.

Liu X, Brutlag DL, Liu JS. (2001) BioProspector: Discovering conserved DNA motifs in upstream regulatory regions of co-expressed genes. *Pac Symp Biocomput*, 127–138.

Liu XS, Brutlag DL, Liu JS. (2002) An algorithm for finding protein-DNA binding sites with applications to chromatin-immunoprecipitation microarray experiments. *Nat Biotechnol* **20**: 835–839.

Loh YH, Wu Q, Chew JL *et al.* (2006) The Oct4 and Nanog transcription network regulates pluripotency in mouse embryonic stem cells. *Nat Genet* **38**: 431–440.

Marchal K, Thijs G, De Keersmaecker S, Monsieurs P, De Moor B, Vanderleyden J. (2003) Genome-specific higher-order background models to improve motif detection. *Trends Microbiol* **11**: 61–66.

Mardis ER. (2007) ChIP-seq: Welcome to the new frontier. *Nat Methods* **4**: 613–614.

Mardis ER. (2008) The impact of next-generation sequencing technology on genetics. *Trends Genet* **24**: 133–141.

Marsan L, Sagot MF. (2000) Algorithms for extracting structured motifs using a suffix tree with an application to promoter and regulatory site consensus identification. *J Comput Biol* **7**: 345–362.

Marschall T, Rahmann S. (2009) Efficient exact motif discovery. *Bioinformatics* **25**: i356–364.

Narasimhan C, LoCascio P, Uberbacher E. (2003) Background rareness-based iterative multiple sequence alignment algorithm for regulatory element detection. *Bioinformatics* **19**: 1952–1963.

Narlikar L, Gordan R, Hartemink AJ. (2007) A nucleosome-guided map of transcription factor binding sites in yeast. *PLoS Comput Biol* **3**: e215.

Neuwald AF, Liu JS, Lawrence CE. (1995) Gibbs motif sampling: Detection of bacterial outer membrane protein repeats. *Protein Sci* **4**: 1618–1632.

Nomenclature Committee of the International Union of Biochemistry (NC-IUB). (1986) Nomenclature for incompletely specified bases in nucleic acid sequences. Recommendations 1984. *Proc Natl Acad Sci USA* **83**: 4–8.

Odom DT, Dowell RD, Jacobsen ES *et al.* (2007) Tissue-specific transcriptional regulation has diverged significantly between human and mouse. *Nat Genet* **39**: 730–732.

Park PJ. (2009) ChIP-seq: Advantages and challenges of a maturing technology. *Nat Rev Genet* **10**(10): 669–680.

Pavesi G, Mauri G, Pesole G. (2001) An algorithm for finding signals of unknown length in DNA sequences. *Bioinformatics* **17** **Suppl 1**: S207–214.

Pavesi G, Mauri G, Pesole G. (2004) *In silico* representation and discovery of transcription factor binding sites. *Brief Bioinform* **5**: 217–236.

Pavesi G, Mereghetti, P, Mauri G, Pesole G. (2004) Weeder Web: Discovery of transcription factor binding sites in a set of sequences from co-regulated genes. *Nucleic Acids Res* **32**: W199–203.

Pevzner PA, Sze SH. (2000) Combinatorial approaches to finding subtle signals in DNA sequences. *Proc Int Conf Intell Syst Mol Biol* **8**: 269–278.

Pillai S, Chellappan SP. (2009) ChIP on chip assays: Genome-wide analysis of transcription factor binding and histone modifications. *Methods Mol Biol* **523**: 341–366.

Roven C, Bussemaker HJ. (2003) REDUCE: An online tool for inferring cis-regulatory elements and transcriptional module activities from microarray data. *Nucleic Acids Res* **31**: 3487–3490.

Sadler JR, Waterman MS, Smith TF. (1983) Regulatory pattern identification in nucleic acid sequences. *Nucleic Acids Res* **11**: 2221–2231.

Sandve GK, Abul O, Walseng V, Drablos F. (2007) Improved benchmarks for computational motif discovery. *BMC Bioinformatics* **8**: 193.

Sandve GK, Drablos F. (2006) A survey of motif discovery methods in an integrated framework. *Biol Direct* **1**: 11.

Sauer T, Shelest E, Wingender E. (2006) Evaluating phylogenetic footprinting for human-rodent comparisons. *Bioinformatics* **22**: 430–437.

Schulze A, Downward J. (2001) Navigating gene expression using microarrays — a technology review. *Nat Cell Biol* **3**: E190–195.

Shinozaki D, Akutsu T, Maruyama O. (2003) Finding optimal degenerate patterns in DNA sequences. *Bioinformatics* **19 Suppl 2**: II206–II214.

Siddharthan R, Siggia ED, van Nimwegen E. (2005) PhyloGibbs: A Gibbs sampling motif finder that incorporates phylogeny. *PLoS Comput Biol* **1**: e67.

Sinha S, Blanchette M, Tompa M. (2004) PhyME: A probabilistic algorithm for finding motifs in sets of orthologous sequences. *BMC Bioinformatics* **5**: 170.

Sinha S, Tompa M. (2003) YMF: A program for discovery of novel transcription factor binding sites by statistical overrepresentation. *Nucleic Acids Res* **31**: 3586–3588.

Sinha S, Tompa M. (2003) *Third IEEE Symposium on Bioinformatics and Bioengineering*, Washington DC, pp. 69–78.

Stormo GD. (2000) DNA binding sites: Representation and discovery. *Bioinformatics* **16**: 16–23.

Sze SH, Gelfand MS, Pevzner PA. (2002) Finding weak motifs in DNA sequences. *Pac Symp Biocomput*, 235–246.

Tang MH, Krogh A, Winther O. (2008) BayesMD: Flexible biological modeling for motif discovery. *J Comput Biol* **15**: 1347–1363.

Thijs G, Lescot M, Marchal K *et al.* (2001) A higher-order background model improves the detection of promoter regulatory elements by Gibbs sampling. *Bioinformatics* **17**: 1113–1122.

Thijs G, Marchal K, Lescot M *et al.* (2002) A Gibbs sampling method to detect overrepresented motifs in the upstream regions of coexpressed genes. *J Comput Biol* **9**: 447–464.

Thompson W, Rouchka EC, Lawrence CE. (2003) Gibbs Recursive Sampler: Finding transcription factor binding sites. *Nucleic Acids Res* **31**: 3580–3585.

Tompa M, Li N, Bailey TL *et al.* (2005) Assessing computational tools for the discovery of transcription factor binding sites. *Nat Biotechnol* **23**: 137–144.

van Helden J, Andre B, Collado-Vides J. (1998) Extracting regulatory sites from the upstream region of yeast genes by computational analysis of oligonucleotide frequencies. *J Mol Biol* **281**: 827–842.

Waterman MS, Arratia R, Galas DJ. (1984) Pattern recognition in several sequences: Consensus and alignment. *Bull Math Biol* **46**: 515–527.

Workman CT, Stormo GD. (2000) ANN-Spec: A method for discovering transcription factor binding sites with improved specificity. *Pac Symp Biocomput* 467–478.

Zeller KI, Zhao X, Lee CW *et al.* (2006) Global mapping of c-Myc binding sites and target gene networks in human B cells. *Proc Natl Acad Sci USA* **103**: 17834–17839.

Chapter 5

A New Approach to the Discovery
of RNA Structural Elements
in the Human Genome

Lei Hua*, Miguel Cervantes-Cervantes[†]
and Jason T. L. Wang*

Analysis of a large number of RNA molecules indicates that variations in their nucleotide sequences do not necessarily convey differences in their secondary structures. Numerous methods have been developed to find patterns in RNA molecules, including the detection of structural motifs in families of noncoding RNAs (ncRNAs). When almost identical sequences render similar structures, these methods work well. However, when given structures differ from each other, there may not exist a motif common to all of them. In such cases, it is desirable to find out if such motifs are indeed present and if not, to determine the extent to which they are shared by the structures under study. We present here a novel tool to be used in finding common patterns among RNAs. In particular, we describe the use of this tool to find RNA structural elements in the human genome. Many of the RNA structures found by our method overlap with human genomic regions that have been previously found through other genome-wide studies aimed to discover conserved structured RNAs. Our method thus provides a complementary tool to the currently used approaches for mining conserved structured RNAs in the human genome.

1. Introduction

In addition to its central role in translation of a cell's genetic information into proteins, ribonucleic acid (RNA) has several other functions, a number

*Department of Computer Science, New Jersey Institute of Technology, Newark, NJ 07102, USA
[†]Department of Biological Sciences, Rutgers University, Newark, NJ 07102, USA

of which are attributable to its structural particularities (herein called RNA motifs). RNA molecules other than messenger RNA (mRNA) are designated noncoding RNAs (ncRNAs) and have been extensively studied for the presence of RNA motifs. ncRNAs include transfer RNA (tRNA), ribosomal RNA (rRNA), small nuclear RNA (snRNA), and small nucleolar RNA (snoRNA) (Griffiths-Jones *et al.*, 2003). More recently, small interfering RNA (siRNA) and microRNA (miRNA) have been under intensive analysis (Ambros *et al.*, 2003). Nevertheless, secondary structures in the untranslated regions (UTRs) of messenger RNAs (mRNAs) have not been characterized in-depth (Pesole *et al.*, 2002). The importance of studying UTRs emerges from a myriad of biochemical and genetic studies that have demonstrated functions associated with UTRs in mRNA metabolism, including RNA translocation, translation, and maintenance of RNA stability.

Whereas RNA structure determination via biochemical and biophysical experiments is laborious and costly, predictive approaches are valuable in providing guidance for wet lab experiments. RNA structure prediction is usually based on thermodynamics of RNA folding or phylogenetic conservation of base-paired regions. The former uses thermodynamic properties of various RNA local structures, such as base-pair stacking, hairpin loop, and bulge, to derive thermodynamically favorable secondary structures. Optimal or suboptimal structures can be found by using dynamic programming algorithms such as the well-known tools MFOLD (Zuker, 1989) and RNAfold in the Vienna RNA package (Schuster *et al.*, 1994; Hofacker, 2003). Similar tools exist to predict higher-order structures, e.g. pseudoknots (Rivas and Eddy, 1999). On the other hand, RNA structure prediction using phylogenetic information helps in inferring RNA structures based on covariation of based-paired nucleotides (Gulko and Haussler, 1996; Hofacker *et al.*, 2002; Knudsen and Hein, 2003). It is generally believed that methods using phylogenetic information are more accurate but their performance critically depends on high-quality alignment of a large number of structurally related sequences.

Tools utilized in the alignment of biological sequences (DNA, protein), such as FASTA and BLAST, are valuable in identifying homologous regions, which in turn may lead to the discovery of functional units, such as protein domains, DNA *cis* elements and others (Pearson and Lipman, 1988; Altschul *et al.*, 1990). Bioinformaticians are mostly satisfied with these tools for proteins, but not for RNAs. Eddy (Eddy, 2006) pointed out some reasons in his excellent paper presented in the 2006 Cold Spring Harbor Symposium on Quantitative Biology. In the case of proteins, for the task of similarity searching, BLAST is able to identify significant homologies down to 20–30% amino acid sequence identity. Many proteins are conserved at this level across billions of years of divergence. In contrast, significant nucleic acid sequence alignments are only detected down to about 60–70% nucleotide sequence identity, largely due to the smaller nucleotide alphabet. Many conserved RNAs diverge below a 60–70% identity threshold in just tens or hundreds of millions of years. BLAST comparisons of RNAs are thus unable to detect important evolutionary divergences. Therefore, it is necessary to take into account structural information in analyzing RNA data.

We present here a method, named DiscoverR, for identifying common patterns of two RNA secondary structures based on our previous work on tree pattern finding (Wang *et al.*, 1998). DiscoverR works by representing RNA secondary structures as ordered labeled trees and performs tree pattern discovery by allowing certain subtrees to be removed at no cost. This method is able to identify common patterns in the secondary structures of two RNA molecules R_1 and R_2, where the common patterns may contain noncontiguous subsequences in R_1 and R_2. The method can detect distant bases interacting remotely that are part of the common patterns of R_1 and R_2. Both the time complexity and the space complexity of the DiscoverR method is $O(|R_1||R_2|)$ where $|\cdot|$ represents the number of nucleotides in the indicated RNA molecule.

2. Related Work

Several methods have been developed that carry out RNA secondary structure prediction and comparison at the same time. For example, Sankoff's method involves simultaneous folding and aligning of two RNA sequences in $O(N^6)$ time where N is the average length of the sequences (Sankoff, 1985). FOLDALIGN (Gorodkin *et al.*, 2001; Havgaard *et al.*, 2005a) improves Sankoff's method with a time complexity of $O(N^4)$; a faster version of the tool was recently presented in (Havgaard *et al.*, 2007). Havgaard *et al.* (Havgaard *et al.*, 2005b) described a method based on FOLDALIGN and the Sankoff algorithm that is effective for sequences with low similarity, specifically with similarity <40%. Mathews and Turner (Mathews and Turner, 2002) presented Dynalign that runs in time $O(d^3 N^3)$ by restricting the maximum distance allowed, d, between aligned nucleotides in two RNA molecules. Tabei *et al.* (Tabei *et al.*, 2006) aligned RNA sequences by matching fixed-length stem fragments in a very efficient way, and implemented their algorithms into a tool called SCARNA. By taking into account local similarity, stem energy and covariations, Perriquet *et al.* (Perriquet *et al.*, 2003) proposed CARNAC for folding and finding the common structure of two RNA sequences. The theoretical time complexity of CARNAC is $O(N^6)$, which could be reduced to $O(N^2)$ by pre-processing the sequences.

Tools that perform global or local alignment of two given RNA secondary structures include RNAdistance (Shapiro and Zhang, 1990), rna_align (Jiang *et al.*, 2002), RNAforester (Hochsmann *et al.*, 2004) and RSmatch (Liu *et al.*, 2005). RNAdistance (Shapiro and Zhang, 1990) uses a tree-based model to coarsely represent RNA secondary structures, and compares the secondary structures based on the edit distance of trees. The rna_align method (Jiang *et al.*, 2002) models RNA secondary structures by nested and/or crossing arcs that connect bonded nucleotides and aligns the RNA secondary structures efficiently. The time complexity of rna_align is $\min\{O(MN^3), O(M^3 N)\}$ where M, N are the lengths of the input structures. RNAforester (Hochsmann *et al.*, 2004) performs RNA alignment by extending the tree model to a forest model. More recently, RSmatch (Liu *et al.*, 2005; Khaladkar *et al.*, 2007; Khaladkar *et al.*, 2008) adopts a loop model for representing and aligning RNA secondary structures with a time complexity of $O(MN)$. The structural RNA alignment can be displayed and viewed using RNALogo (Chang *et al.*, 2008).

DiscoverR is *not* an alignment method. The tool differs from a global alignment (GA) algorithm in the following way. When globally aligning a small RNA molecule with a large RNA molecule, the GA algorithm inserts many gaps, which would lead to a meaningless alignment result. In this situation, DiscoverR is able to extract common patterns from the two RNA molecules having different sizes without inserting gaps. On the other hand, DiscoverR differs from a local alignment (LA) algorithm in the following way. When locally aligning two RNA molecules, the LA algorithm seeks small, local regions with high similarity where bases are close to each other (Backofen *et al.*, 2007). In contrast, DiscoverR looks at the entire RNA molecules to extract their largest common substructures possibly with distant bases on the respective molecules.

3. Methods

RNA molecules acquire their secondary structures through proper folding (Fig. 1(a)). Let RS be an RNA sequence consisting of nucleotides or bases A, U, C, G. $RS[i]$ denotes the base at position i of RS and $RS[i,j]$ is the subsequence starting at position i and ending at position j in RS. Let R be the secondary structure of RS. A base pair between position i and position j in R is denoted by (i,j) and its enclosed sequence is $RS[i,j]$. A loop in R refers to a hairpin loop H, a bulge loop B, an internal loop I or a multibranched loop M (Mathews *et al.*, 1999; Hofacker, 2003; Zuker, 2003). Given a loop L in the secondary structure R, the base pair (i^*, j^*) in L is called the exterior pair of L if position i^* (j^*, respectively) is closest to the 5′ (3′, respectively) end of R among all positions in L. All other nonexterior base pairs in L are called interior pairs of L(Spirollari *et al.*, 2009).

As previously described by our group and other workers (Shapiro and Zhang, 1990; Hochsmann *et al.*, 2004; Liu *et al.*, 2005; Nawrocki *et al.*, 2009), we modeled the RNA secondary structure R of the sequence RS by a rooted ordered labeled tree RT. The tree has a root, each node has a label and the left-to-right order of siblings is significant. Pseudoknots are not allowed in this model. Each node in RT corresponds to a base pair in R and vice versa. Base pairs are orderly numbered from the 5′-end to the 3′-end of R. Except for the exterior pairs of loops, the kth base pair of R corresponds to the node labeled "Pk" in RT and vice versa. For example, the node labeled "P3" in the tree RT shown in Fig. 1(b) corresponds to the 3rd base pair in the secondary structure R shown in Fig. 1(a). The exterior pair of a multibranched loop containing n interior pairs in R corresponds to a node v with n children in RT with each child corresponding to one of the n interior pairs. Assuming the exterior pair is the kth base pair in R, the node label of v is "Mk". The exterior pair of a bulge loop (internal loop, hairpin loop, respectively) in R corresponds to the node labeled "Bk" ("Ik", "Hk", respectively) in RT if the exterior pair is the kth base pair in R. For example, the node labeled "M5" ("B18", "I24", "H31", respectively) in the tree RT shown in Fig. 1(b) corresponds to the exterior pair of the multibranched loop (bulge loop, internal loop, hairpin loop, respectively) where the exterior pair is the 5th (18th, 24th, 31st, respectively) base pair in the RNA secondary structure R shown in Fig. 1(a). For each node v in the tree RT, we use $NB(v)$ to represent the

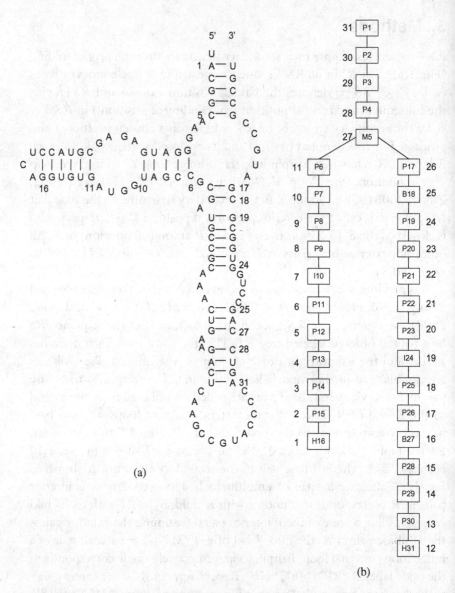

(a)

(b)

Fig. 1. The tree representation of RNA secondary structure. (a) An RNA secondary structure is comprised of base pairs, which are numbered according to the order from the 5′ end to the 3′ end of the secondary structure. (b) The base pairs are organized into an ordered labeled tree. Each node in the tree corresponds to a base pair in the secondary structure and vice versa. The numeric number next to each node in the tree is the position of that node in the left-to-right postorder traversal of the tree.

number of bases v has. If the node label of v is "Pi" for some i, $NB(v) = 2$. If v corresponds to the exterior pair of a loop, $NB(v)$ equals the number of bases in that loop.

Let R_1 and R_2 be two RNA secondary structures. Let RT_1 (RT_2, respectively) be the tree representing R_1 (R_2, respectively). Let rt_1 be a node in RT_1 and let rt_2 be a node in RT_2. The dissimilarity between the two nodes rt_1 and rt_2, denoted $\delta(rt_1, rt_2)$, is calculated as follows:

$$\delta(rt_1, rt_2) = \frac{|NB(rt_1) - NB(rt_2)|}{|NB(rt_1) + NB(rt_2)|}.$$

Thus, $\delta(rt_1, rt_2)$ equals 0 if rt_1 and rt_2 have the same number of bases. We say node rt_1 matches node rt_2, denoted $rt_1 \approx rt_2$, if $\delta(rt_1, rt_2) \leq \varepsilon$, where ε is a user-determined threshold. We say tree RT_1 matches tree RT_2, denoted $RT_1 \approx RT_2$, if the two trees are isomorphic and each node in RT_1 matches its corresponding node in RT_2. DiscoverR is based on the dynamic programming algorithm presented in (Wang *et al.*, 1998), which is capable of finding the largest common structures of two trees.

The DiscoverR program is available on the web at `http://datalab.njit.edu/rna/DiscoverR/`, which can find the largest common substructures or patterns of two RNA secondary structures modeled by trees as described above. Figure 2 shows the output of the DiscoverR program given two RNA secondary structures as input, where beginning and ending positions of the contiguous bases on the common patterns in the two input structures are printed out. In Fig. 3, the output of the DiscoverR program is portrayed using RnaViz 2 (Rijk *et al.*, 2003), where the common patterns of the two input structures are highlighted. This tool is used to mine conserved structured RNAs in the human genome, as described in the next section.

4. Results

We applied DiscoverR to finding conserved RNA secondary structures in the human genome and examined how the structures we found differ from the results obtained from other studies that were recently carried out for finding conserved RNA secondary structures in the human genome (Washietl *et al.*, 2005; Pedersen *et al.*, 2006; Khaladkar *et al.*, 2008). Using 8-way human-referenced vertebrate genome alignments, Washietl *et al.* (Washietl *et al.*, 2005) detected 91 676 conserved RNA

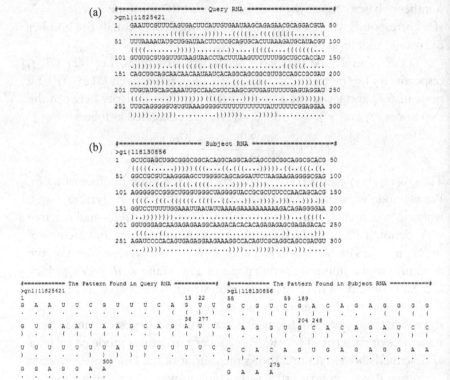

Fig. 2. (a) A query RNA. (b) A subject RNA. (c) The pattern found in the query RNA. (d) The pattern found in the subject RNA. In (c), (d), beginning and ending positions of the contiguous bases on the common patterns found in the query RNA and subject RNA are displayed.

structures (at $p > 0.5$) using the RNAz program, which identified RNA structures with similar thermodynamic stabilities across multiple species. Pedersen *et al.* (Pedersen *et al.*, 2006) developed a phylogenetics-based stochastic context-free grammar (phylo-SCFG), and identified 48 479 candidate RNA structures using the same genome alignments. Torarinsson *et al.* (Torarinsson *et al.*, 2006) focused on human and mouse genomic sequences that could not be aligned on the sequence level, and identified conserved structures by FOLDALIGN surveyed in the Related Work section. Khaladkar *et al.* (Khaladkar *et al.*, 2008) developed a clustering-based approach, named GLEAN-UTR, to identify stem-loop RNA structure

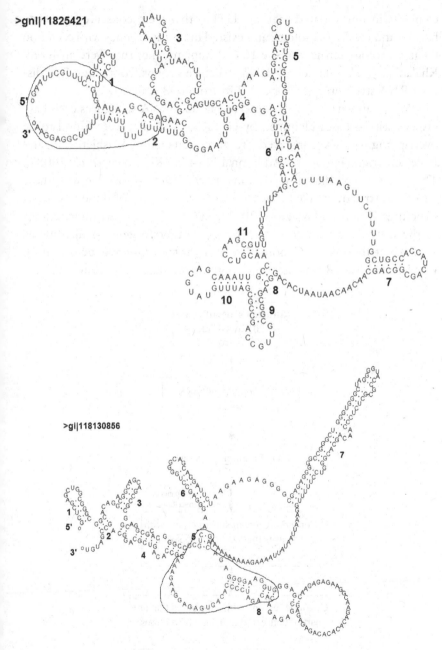

Fig. 3. Common patterns found by DiscoverR in the two RNA molecules gnl|11825421 and gi|118130856 in Fig. 2 are highlighted.

elements in untranslated regions (UTRs) that were conserved between human and mouse orthologs, and existed in multiple genes with common Gene Ontology terms. For the 10 448 human genes that were analyzed, Khaladkar *et al.* obtained 90 RNA structure groups, containing 748 distinct RNA structures in 5′ or 3′ UTRs from 698 genes.

We began with the 130 conserved human RNA structures each having at least 14 bases identified by GLEAN-UTR that were found to be overlapping with the conserved structures detected by Washietl *et al.* and Pedersen *et al.* (Fig. 4 and Additional file 4 in (Khaladkar *et al.*, 2008)). The structures predicted by Torarinsson *et al.* did not overlap with these 130 RNA structures. We located the genomic regions of these 130 RNA structures (Pruitt and Maglott, 2001), and mapped the genomic regions to the 8-way human-referenced (hg17) vertebrate genome alignments available at the UCSC Genome Browser (http://genome.ucsc.edu/). We selected the 8-way genome alignments that fully contained the

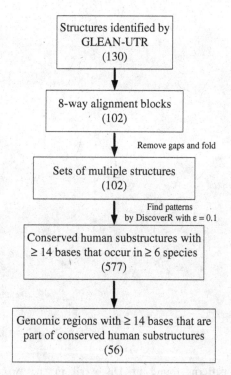

Fig. 4. Illustration of the flowchart of our approach for mining conserved structured RNAs in the human genome.

genomic regions of the RNA structures (if a structure straddled two different genome alignments, that structure was excluded). Some of the selected genome alignments were long, with several thousand nucleotides. We extracted a subalignment or alignment block from each selected genome alignment where the length of an alignment block was Ln and each alignment block fully contained the genomic region of at least one structure listed in Additional file 4 in (Khaladkar *et al.*, 2008). (In the study presented here, Ln was set to 300.) If the length of a selected genome alignment was less than Ln, that whole genome alignment was treated as an alignment block. This step resulted in 102 alignment blocks where each alignment block had four to eight sequences (species).

We then designed a systematic approach to detecting conserved human structures using DiscoverR, which worked as follows. For each alignment block B, we removed gaps in it and obtained a set S_B of eight or fewer sequences for that alignment block. Using the Vienna RNA Package (version 1.7.2) (Hofacker, 2003), we folded each sequence in S_B to get its minimum-energy secondary structure, also placed in S_B. We then compared the human structure, H, with each of the other structures, R, in S_B using Discover R(with $\varepsilon = 0.1$). Specifically, for the tree HT representing H and the tree RT representing R, we found the largest common substructures of $HT[i]$ and $RT[j]$, for all $1 \le i \le |HT|$ and $1 \le j \le |RT|$. The found patterns or substructures of the human structure H were stored in a list, denoted *List*. Each substructure in *List* had at least 14 bases as in (Khaladkar *et al.*, 2008); substructures with less than 14 bases were excluded from *List*. We identified those human substructures in *List* that occurred in at least *Occur* secondary structures in S_B. (In the study presented here, *Occur* was set to 6.) If the number of secondary structures in S_B was less than *Occur*, no substructure in *List* qualified to be a solution. Here a solution was a conserved human substructure that occurred in at least *Occur* species and had at least 14 bases. This step resulted in 577 qualified substructures. Among the 577 found substructures, some were substructures of others; these subpatterns were eliminated from further consideration. Within the remaining qualified substructures, there were 56 genomic regions each having at least 14 contiguous bases (short regions with less than 14 bases were not considered as in (Khaladkar *et al.*, 2008)). This structure mining method is illustrated in Fig. 4. The genomic regions within the conserved human substructures found by our approach are listed in Table 1. It can be seen from Table 1 that some of the conserved

Table 1.　Results of the experiments performed in this study.

Chromosome	Start Position	Length	Strand	Overlap with
Genomic regions within conserved human substructures that occur in eight species.				
Chr6	26265302	14	+	P
ChrX	106763864	27	–	K, P
Genomic regions within conserved human substructures that occur in seven species.				
Chr1	96992213	19	+	W
Chr2	144979616	17	–	–
Chr2	144979712	17	–	K, P
Chr3	197265646	23	–	P
Chr3	197265693	23	–	–
Chr3	197265758	23	–	K, P
Chr3	197266161	25	–	K, P
Chr3	197266211	25	–	K, P
Chr6	19947838	20	+	K, W, P
Chr6	19947871	21	+	K, W
Chr6	168884945	15	+	K, P
Chr14	28308288	16	+	–
Chr17	59926225	18	–	–
ChrX	106763863	29	–	K, P
Genomic regions within conserved human substructures that occur in six species.				
Chr1	8006261	15	–	–
Chr1	96991635	15	+	W
Chr1	96991697	18	+	W
Chr1	96992209	27	+	K, W
Chr2	14726767	16	+	W
Chr2	144979653	18	–	P
Chr2	144979710	21	–	K, P
Chr2	144979850	16	–	–
Chr2	190270896	21	–	K, P
Chr3	37835407	14	+	–
Chr3	37835524	14	+	–
Chr3	161701103	17	–	K, P
Chr3	161701244	17	–	P
Chr3	197265645	25	–	P
Chr3	197265757	25	–	K, P
Chr3	197266160	27	–	K, P

(*Continued*)

Table 1. (*Continued*)

Chromosome	Start Position	Length	Strand	Overlap with
Genomic regions within conserved human substructures that occur in six species.				
Chr3	197266210	27	−	K, P
Chr5	179136573	17	+	P, W
Chr5	179136607	17	+	W
Chr5	179136709	16	+	−
Chr6	134532540	16	−	K, P
Chr7	38196970	16	−	K, P
Chr7	77229459	15	+	W
Chr7	77229519	33	+	K, P, W
Chr7	101486144	16	+	K, P
Chr7	101486165	14	+	W
Chr8	117927555	16	−	−
Chr8	117927581	22	−	−
Chr8	136728826	16	+	W
Chr9	89160953	15	+	K, P, W
Chr10	30790413	19	+	K, W, P
Chr10	119298963	14	+	−
Chr10	119298984	17	+	P
Chr14	28308435	17	+	W, P
Chr14	53964017	14	−	−
Chr14	53964115	22	−	−
Chr17	59926370	23	−	−
Chr19	1386284	25	+	P
Chr19	39410717	38	+	K, P, W
Chr19	39410811	20	+	W

human substructures found by our approach overlap with the known structures detected by the existing methods (K for GLEAN-UTR (Khaladkar *et al.*, 2008), P for Pedersen *et al.* (Pedersen *et al.*, 2006) and W for Washietl *et al.* (Washietl *et al.*, 2005)), while others are novel ones that are not identified previously.

5. Conclusion

We presented a new tool (DiscoverR) for finding common patterns in RNAs and described a new approach to detecting conserved human structures using DiscoverR. Experimental results showed that the proposed approach is able to locate many genomic regions with conserved RNA secondary structures where some of the genomic regions overlap with known regions detected by the existing methods (Washietl *et al.*, 2005; Pedersen *et al.*, 2006; Khaladkar *et al.*, 2008) while others are not reported previously. These results demonstrate that DiscoverR is a useful tool for RNA motif discovery (Leontis and Westhof, 2002; 2003). The findings also indicate that there may exist much more conserved RNA secondary structures in the human genome that remain to be explored.

References

Altschul SF, Gish W, Miller W, Myers EW, Lipman DJ. (1990) Basic local alignment search tool. *J Mol Biol* **215**: 403–410.

Ambros V, Bartel B, Bartel DP *et al.* (2003) A uniform system for microRNA annotation. *RNA* **9**: 277–279.

Backofen R, Chen S, Hermelin D *et al.* (2007) Locality and gaps in RNA comparison. *J Comput Biol* **14**: 1074–1087.

Chang TH, Horng JT, Huang HD. (2008) RNALogo: A new approach to display structural RNA alignment. *Nucleic Acids Res* **36**: W91–W96.

Eddy SR. (2006) Computational analysis of RNAs. *Proceedings of the 71st Cold Spring Harbor Symposium on Quantitative Biology* **71**: 117–128.

Gorodkin J, Stricklin SL, Stormo GD. (2001) Discovering common stem loop motifs in unaligned RNA sequences. *Nucleic Acids Res* **29**: 2135–2144.

Griffiths-Jones S, Bateman A, Marshall M, Khanna A, Eddy SR. (2003) Rfam: An RNA family database. *Nucleic Acids Res* **31**: 439–441.

Gulko B, Haussler D. (1996) Using multiple alignments and phylogenetic trees to detect RNA secondary structure. *Pac Symp Biocomput*, 350–367.

Havgaard JH, Lyngso RB, Gorodkin J. (2005a) The FOLDALIGN web server for pairwise structural RNA alignment and mutual motif search. *Nucleic Acids Res* **33**: W650–W653.

Havgaard JH, Lyngso RB, Stormo GD, Gorodkin J. (2005b) Pairwise local structural alignment of RNA sequences with sequence similarity less than 40%. *Bioinformatics* **21**: 1815–1824.

Havgaard JH, Torarinsson E, Gorodkin J. (2007) Fast pairwise structural RNA alignments by pruning of the dynamical programming matrix. *PLoS Comput Biol* **3**: e193.

Hochsmann M, Voss B, Giegerich R. (2004) Pure multiple RNA secondary structure alignments: A progressive profile approach. *IEEE/ACM Trans Comput Biol Bioinform* **1**: 53–62.

Hofacker IL. (2003) Vienna RNA secondary structure server. *Nucleic Acids Res* **31**: 3429–3431.

Hofacker IL, Fekete M, Stadler PF. (2002) Secondary structure prediction for aligned RNA sequences. *J Mol Biol* **319**: 1059–1066.

Jiang T, Lin G, Ma B, Zhang K. (2002) A general edit distance between RNA structures. *J Comput Biol* 9: 371–388.

Khaladkar M, Bellofatto V, Wang JTL, Tian B, Shapiro BA. (2007) RADAR: A web server for RNA data analysis and research. *Nucleic Acids Res* 35: W300–W304.

Khaladkar M, Liu J, Wen D, Wang JTL, Tian B. (2008) Mining small RNA structure elements in untranslated regions of human and mouse mRNAs using structure-based alignment. *BMC Genomics* 9: 189.

Knudsen B, Hein J. (2003) Pfold: RNA secondary structure prediction using stochastic context-free grammars. *Nucleic Acids Res* 31: 3423–3428.

Leontis NB, Westhof E. (2002) The annotation of RNA motifs. *Comparative and Functional Genomics* 3: 518–524.

Leontis NB, Westhof E. (2003) Analysis of RNA motifs. *Curr Opin Struct Biol* 13: 300–308.

Liu J, Wang JTL, Hu J, Tian B. (2005) A method for aligning RNA secondary structures and its application to RNA motif detection. *BMC Bioinformatics* 6: 89.

Mathews DH, Turner DH. (2002) Dynalign: An algorithm for finding the secondary structure common to two RNA sequences. *J Mol Biol* 317: 191–203.

Mathews DH, Sabina J, Zuker M, Turner DH. (1999) Expanded sequence dependence of thermodynamic parameters improves prediction of RNA secondary structure. *J Mol Biol* 288: 911–944.

Nawrocki EP, Kolbe DL, Eddy SR. (2009) Infernal 1.0: Inference of RNA alignments. *Bioinformatics* 25(10): 1335–1337.

Pearson WR, Lipman DJ. (1988) Improved tools for biological sequence comparison. *Proc Nat Acad Sci USA* 85: 2444–2448.

Pedersen JS, Bejerano G, Siepel A *et al.* (2006) Identification and classification of conserved RNA secondary structures in the human genome. *PLoS Comput Biol* 2(4): e33.

Perriquet O, Touzet H, Dauchet M. (2003) Finding the common structure shared by two homologous RNAs. *Bioinformatics* 19: 108–118.

Pesole G, Liuni S, Grillo G *et al.* (2002) UTRdb and UTRsite: Specialized databases of sequences and functional elements of 5′ and 3′ untranslated regions of eukaryotic mRNAs. *Nucleic Acids Res* 30: 335–340.

Pruitt KD, Maglott DR. (2001) RefSeq and LocusLink: NCBI gene-centered resources. *Nucleic Acids Res* 29: 137–140.

Rijk PD, Wuyts J, Wachter RD. (2003) RnaViz2: An improved representation of RNA secondary structure. *Bioinformatics* 19: 299–300.

Rivas E, Eddy SR. (1999) A dynamic programming algorithm for RNA structure prediction including pseudoknots. *J Mol Biol* 285: 2053–2068.

Sankoff D. (1985) Simultaneous solution of the RNA folding, alignment and proto-sequence problems. *SIAM J Appl Math* 45: 810–825.

Schuster P, Fontana W, Stadler PF, Hofacker IL. (1994) From sequences to shapes and back: A case study in RNA secondary structures. *Proc Biol Sci* 255: 279–284.

Shapiro BA, Zhang K. (1990) Comparing multiple RNA secondary structures using tree comparisons. *Computer Appl Biosci* 6: 309–318.

Spirollari J, Wang JTL, Zhang K, Bellofatto V, Park Y, Shapiro BA. (2009) Predicting consensus structures for RNA alignments via pseudo-energy minimization. *Bioinform Biol Insights* 3: 51–69.

Tabei Y, Tsuda K, Kin T, Asai K. (2006) SCARNA: Fast and accurate structural alignment of RNA sequences by matching fixed-length stem fragments. *Bioinformatics*
22: 1723–1729.

Torarinsson E, Sawera M, Havgaard JH, Fredholm M, Gorodkin J. (2006) Thousands of corresponding human and mouse genomic regions unalignable in primary sequence contain common RNA structure. *Genome Res* **16**: 885–889.

Wang JTL, Shapiro BA, Shasha D, Zhang K, Currey KM. (1998) An algorithm for
finding the largest approximately common substructures of two trees. *IEEE Trans
Pattern Anal Mach Intell* **20**: 889–895.

Washietl S, Hofacker IL, Lukasser M, Huttenhofer A, Stadler PF. (2005) Mapping of
conserved RNA secondary structures predicts thousands of functional noncoding
RNAs in the human genome. *Nat Biotechnol* **23**(11): 1383–1390.

Zuker M. (1989) Computer prediction of RNA structure. *Methods Enzymol* **180**:
262–288.

Zuker M. (2003) Mfold web server for nucleic acid folding and hybridization
prediction. *Nucleic Acids Res* **31**: 3406–3415.

Part II

PERFORMANCE AND PARADIGMS

Chapter 6

Benchmarking of Methods
for Motif Discovery in DNA

Kjetil Klepper*, Geir Kjetil Sandve†, Morten Beck Rye*,
Kjersti Hysing Bolstad* and Finn Drabløs*,‡

Reliable benchmarking is a prerequisite for unbiased comparison of alternative approaches to motif discovery in DNA. This chapter focuses on datasets and procedures for benchmarking of discovery methods for transcription factor binding sites in genomic DNA. The general concept of benchmarking is discussed briefly, including the importance of using separate training data, test data and benchmark data. The need to distinguish between different aspects of the motif discovery process during benchmarking is also discussed. The various score functions for binding site predictions are presented, including both sequence based scoring and position-weight matrix based scoring as well as the relationship between these. The problem of unannotated binding sites in real sequences as well as the high rate of false positive predictions normally associated with motif discovery is also discussed in this context. Relevant criteria for good benchmark sets and some frequently used datasets are presented, together with new and improved sets for benchmarking of both single motif and module based prediction methods. Approaches towards genome-wide benchmarking using ChIP-chip and in particular ChIP-seq data are evaluated, including the problem of identifying functionally significant weak binding sites and benchmarking of methods using additional genome data as prior information. This is extended into a discussion of benchmarking without access to high-quality reference datasets of known binding sites. Finally, benchmarking of prediction methods of particular relevance to genome-based motif discovery is briefly discussed, including promoter and transcription start site prediction and nucleosome positioning.

---

*Department of Cancer Research and Molecular Medicine, Norwegian University of Science and Technology (NTNU), Trondheim, Norway
†Department of Informatics, University of Oslo, Norway
‡finn.drablos@ntnu.no

1. Introduction

It has been demonstrated several times that robust *in silico* identification of transcription factor binding sites (TFBSs) in eukaryote genomes is an extremely challenging problem (Tompa *et al.*, 2005). As a consequence, very many different approaches have been tried; in a survey from 2006 more than 100 different published methods were identified (Sandve and Drablos, 2006), and the number is still growing. These methods use different algorithms and motif representations, and many methods also include additional information like evolutionary sequence conservation or interaction between binding sites into the search process. However, the improvement in performance is in most cases only marginal. It is therefore important to ensure that the criteria used for comparing different methods are relevant to the problem at hand, sensitive to significant differences between methods, and generally accepted within the research community. It is also important to consider alternative benchmarking strategies and datasets that may be needed for proper evaluation of novel approaches to motif discovery.

A benchmark may be defined as a standard by which something is evaluated or measured. The term originates from surveying, where a chiseled mark in a rock was used to indicate the position of the bench for the levelling rod. A benchmark was therefore a fixed reference point for repeated measurements. The term is now used in several areas to indicate a reference-based approach for comparing performance. In computer science, specific software applications are used to compare the performance of computer systems, as for example the LINPACK benchmark used to rank supercomputers (http://www.top500.org/). In a similar way, specific datasets can be used to compare the performance of software tools.

In this paper, we focus on benchmarking of methods for identification of transcription factor binding sites in DNA. In particular, we explore relevant score functions and suitable benchmark datasets, and indicate best practice with respect to actual benchmarking. However, it is a problem that many datasets seem to be under-annotated (Hawkins *et al.*, 2009), leading to misclassification of predictions as false positives. We will therefore look at improved datasets for benchmarking and also discuss briefly the potential for benchmarking without having to rely on annotation data for the benchmark dataset.

2. Score Functions

A typical benchmarking study includes one or more sequence sets with annotated binding sites (often described as the "gold standard") and one or more score functions for evaluating the match between predicted and known binding sites. Most score functions have been developed for this situation. However, many of these functions are strongly affected by the high rate of false positives that are normally encountered in motif discovery. This situation is often described by what is known as "the futility theorem" (Wasserman and Sandelin, 2004). It is therefore worth considering alternative score functions, or even scoring where no information about known binding sites is used at all, as discussed towards the end of this chapter.

2.1. *Scoring by known binding sites*

Based on the transcription factor (TF) binding site locations predicted by a motif discovery tool, each nucleotide in the sequence can be labeled as being either in a predicted binding site or in a predicted background. To evaluate the performance of a program, its predictions can be compared against the location of known motifs, using real sites in real background sequence or sites inserted into synthetic background sequences.

When a program correctly predicts that a nucleotide is part of a binding site, this is called a *true positive* prediction (TP). A nucleotide correctly predicted as not being part of a binding site is a *true negative* (TN). If a nucleotide is predicted as lying within a binding site when in reality it is not, this is called a *false positive* prediction (FP), also known as a Type I error. A missed binding site position on the other hand is called a *false negative* (FN) or Type II error. The number of TP, TN, FP and FN predictions (Fig. 1) serves as basis for several integrated performance measures, the formulas for which are listed in Table 1.

Sensitivity (Sn, also called *recall rate*) is a measure of the fraction of true binding site nucleotides that have been correctly discovered by a program. It is found by dividing the true positive predictions by the total number of true sites (the ones correctly predicted (TP) and the ones missed (FN)). A program that has high sensitivity is less likely to miss out on true binding sites (or commit type II errors). This measure should never be interpreted in isolation, however, since it is possible for a method to obtain

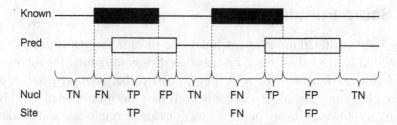

Fig. 1. Comparing known and predicted binding sites. Known and predicted (Pred) binding sites in a DNA sequence are compared at the level of nucleotides (Nucl) and individual binding sites (Site) and classified into TP, TN, FP and FN nucleotides or TP, FP or FN sites. The classification of sites depends on the degree of overlap between known and predicted sites, where e.g. an overlap of at least 25% is required for a positive prediction.

Table 1.   Formulas for commonly used score functions.

Measure	Formula	Range
Sensitivity	$Sn = TP/(TP + FN)$	0 to 1
Specificity	$Sp = TN/(TN + FP)$	0 to 1
Positive Predictive Value	$PPV = TP/(TP + FP)$	0 to 1
Negative Predictive Value	$NPV = TN/(TN + FN)$	0 to 1
Performance Coefficient	$PC = TP/(TP + FN + FP)$	0 to 1
Average Performance	$AP = (Sn + PPV)/2$	0 to 1
F-measure	$F = 2 \times (Sn \times PPV)/(Sn + PPV)$	0 to 1
Accuracy	$Acc = (TP + TN)/(TP + TN + FP + FN)$	0 to 1
Correlation Coefficient	$CC = \frac{TP \times TN - FN \times FP}{(TP+FN)(TN+FP)(TP+FP)(TN+FN)}$	-1 to 1

Average Performance at the site level is often known as Average Site Performance (ASP), and Average Nucleotide Performance (ANP) may be used in an analogous way. For other measures n and s are often used to distinguish between nucleotide or site level statistics (e.g. $nPPV$ or $sPPV$).

a perfect sensitivity score simply by predicting all nucleotides as belonging to binding sites.

Sensitivity is thus usually seen in combination with *specificity* (Sp), which is analogous to sensitivity except that it rates nonbinding site nucleotides. Specificity is the number of nucleotides correctly predicted as background divided by the total number of true background nucleotides.

Programs that show high specificity are less likely to make false binding site predictions (commit type I errors).

Another measure to compare against sensitivity is the *positive predictive value* (PPV) or *precision rate*, which is the proportion of nucleotide binding site predictions made by a program that actually correspond to real binding sites. PPV is sometimes preferred over Sp in motif discovery because it is not biased by the large number of nonbinding sites in these datasets. A high sensitivity score combined with a low PPV would suggest that a program is able to find a high proportion of the true binding sites in a sequence but at the expense of simultaneously making a lot of undesirable false predictions. Analogous to positive predictive value, but not as frequently used for motif discovery evaluation, the negative predictive value (NPV) is the proportion of predicted background nucleotides that are indeed background.

While sensitivity is the fraction of correctly predicted binding sites in relation to all true sites and PPV is the fraction of correctly predicted sites in relation to all predicted sites, the *performance coefficient* (PC) captures aspects of both of these measures. PC is the number of true positive predictions divided by the sum of all true binding sites (TP and FN) and all predicted binding sites (FP). A more straightforward combination of sensitivity and PPV is the *average performance* (AP), which is simply the arithmetic mean of these two measures, while the related F-measure is the harmonic mean. The PC, AP and F measures reward programs that are able to correctly locate more true binding sites while at the same time penalize spurious binding site predictions. For the AP and F-measure, it is also easy to weight the relative importance of precision and recall by slightly modifying the basic formulas.

Each of the measures mentioned so far use only two or three out of the four TP, TN, FP and FN variables. They are thus aimed at measuring particular aspects of performance, and it is often possible for a program to optimize for one or more of these measures at the expense of others. For instance, programs can generally obtain high sensitivity scores by making numerous binding site predictions or high specificity scores by being more conservative.

One measure that combines all four variables is *accuracy*. Accuracy is the fraction of nucleotides that are correctly classified, either as binding site or background. Although this measure provides a more overall view of performance, the result can still be biased if the number of binding site nucleotides in a sequence is skewed compared to the number of background nucleotides, which is usually the case. If for instance the ratio of binding sites to background is small, a high accuracy can be obtained by predicting all nucleotides as background. Another measure that combines all four variables and also accounts for differences in the number of binding site and background nucleotides is the *correlation coefficient* (CC). CC is a measure of the overall agreement between predicted and true locations. A CC score of 1 means that the predictions made by a program are in perfect agreement with the true locations, while a score of -1 would imply the complete opposite: that a program has predicted all true binding sites as background and all background as binding sites. A CC score close to zero would mean that there is no statistical correlation between the predictions made by a program and the location of true binding sites. Such a score is to be expected if the predictions are based on random chance.

Although nucleotide level performance measures are the most accurate, this fine level of evaluation can also be too stringent, since it can severely penalize predictions that are slightly off target even though the prediction itself mostly agrees with the location of a true binding site. This can be problematic, for instance for motif discovery programs that operate with a fixed motif width when the annotated binding sites have varying lengths.

An alternative to nucleotide level scoring is *site level* scoring, where one does not look at whether individual nucleotide positions are predicted correctly or not, but whether the location of a predicted site overlaps with the true binding site. At the site level, a predicted binding site is considered a true positive if there is at least a certain degree of overlap with a true binding site (for instance minimum 25% overlap). A prediction which does not overlap with a true site, or has too little overlap, is considered a false positive prediction, while a true binding site that is not overlapped by any prediction is a false negative. It is not easy to define what we mean by a "site" for negative data, and true negative (TN) predictions are therefore not used at the site level. Hence, the specificity, accuracy and correlation coefficient measures can only be calculated at the nucleotide level.

If the benchmark dataset consists of several sequences, or if several datasets are used for evaluation, it might be desirable to produce a single aggregated score that summarizes the overall performance with respect to a chosen score function. There are different ways to approach this (e.g. Tompa *et al.*, 2005). One option is to calculate the score function independently for each sequence or subset and then average these values to generate a final score. An alternative option is to sum up TP, TN, FP and FN counts for all the sequences and use these combined counts as input to the scoring function. With the first option, the results obtained for each subset will have equal weight in the total score irrespective of the properties of the subset. On the other hand, by combining the counts from all the sequences before calculating the total score, the properties of each subset become important. The net effect depends upon the score function; e.g. some functions are affected by variation in sequence length, but not by variation in motif length, and vice versa.

2.2. *The futility theorem*

The discussion so far has focussed on the comparison between real and predicted binding sites. However, there is a subtle discrepancy between what motif discovery tools actually predict and what we would like them to predict. The "gold standard" reference datasets we use to evaluate methods usually consist of sequences containing experimentally verified functional binding sites, while motif discovery methods look for occurrences of potential motifs in the sequences. The mere presence of a motif, however, does not imply a functional binding site. In fact, it is expected that most binding site motifs encountered in a genome play no functional role whatsoever *in vivo*, even though the sites themselves might very well bind transcription factors *in vitro*. Reasons for this can be that the conformation of chromatin *in vivo* precludes access to certain sites or because transcription factors require the presence of additional factors binding nearby to successfully exert their biological function. It has been estimated that a whole-genome scan with a binding site model for a given transcription factor would typically incur in the order of 1000 false predictions for each functional site (Wasserman and Sandelin, 2004). Thus, predicting binding sites based on sequence similarity alone is essentially a futile undertaking since nearly all sites found this way will have no regulatory role — an assertion which has been termed the "futility theorem." Hence, current motif discovery methods are often plagued with high false-discovery rates, and performance evaluations that are based on comparing predicted sites to annotated sites will consequently suffer.

2.3. *Alternative scoring*

Computational motif discovery is to some extent a two-part process, where motif definition on the one hand and identification of potential binding sites in target sequences on the other hand may be regarded as separate aspects of this process. In most cases, the program either uses a predefined model and finds binding sites by scanning the sequences with that model, or the program does *de novo* motif discovery by optimizing a motif model directly or by optimizing a selection of potential binding sites, which subsequently may be used to build a motif model. In most cases, only the predicted binding sites are evaluated, as already described. However, it may give additional information on benchmark performance if we also evaluate the motif model itself, independent of the specific binding sites, in particular since this evaluation may be less affected by inadequate annotation of binding sites in the benchmark sequences.

A commonly used motif representation is the position weight matrix (PWM) or position specific scoring matrix (PSSM), which is often based on a count matrix from a multiple alignment of binding sites (Wasserman and Sandelin, 2004). Several prediction tools report a count matrix or a PWM as part of the prediction result; otherwise a PWM can easily be generated from the list of predicted binding sites. This PWM will not be strongly affected by unannotated sites as long as they are similar to the known binding sites, and it can be compared to a PWM generated from the benchmark set or from external data (e.g. TRANSFAC (Matys *et al.*, 2003) or Jaspar (Sandelin *et al.*, 2004)), using one of several available methods for comparing matrices (e.g. STAMP (Mahony and Benos, 2007)).

Figure 2 (Bolstad, 2009) shows a comparison between nucleotide level correlation score (nCC) (Tompa *et al.*, 2005) and significance of PWM similarity (*p*-value) (Mahony and Benos, 2007) as performance measures for several *de novo* motif discovery methods (MEME (Bailey *et al.*, 2006), MotifSampler (Thijs *et al.*, 2001), Weeder (Pavesi *et al.*, 2004), AlignACE (Roth *et al.*, 1998) and Ameme (Ao *et al.*, 2004)) tested with alternative parameter settings, using relatively easy ("Algorithm") benchmark sets of random sequences ("Markov") with implanted motifs (Sandve *et al.*, 2007). The figure shows that there is a reasonably clear distinction between successful and unsuccessful cases and a good correlation between the two performance measures, but with a couple of interesting exceptions. These two measures (correlation and PWM similarity) give slightly different and

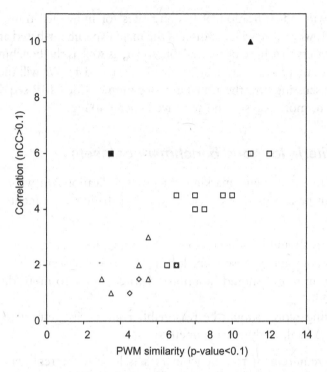

Fig. 2. PWM similarity vs nCC score functions. Performance of several motif discovery tools with alternative parameter settings on a benchmark dataset consisting of real binding sites in randomized background sequences. The tools are plotted according to the number of subsets in the benchmark set with reasonable binding site prediction (nCC (Tompa *et al.*, 2005)) and PWM similarity (*p*-value from STAMP (Mahony and Benos, 2007)). The methods tested were MEME (□), MotifSampler (△), Weeder (■), AlignACE (◇) and Ameme (▲).

potentially complementary information, and there is a clear value in combining information from such alternative performance measures.

3. Benchmark Datasets

It has until now been very challenging to find comprehensive benchmark datasets for motif discovery in DNA. This has been mainly due to limited availability of experimental data with sufficient accuracy. This situation is likely to change with the recent development of chromatin immunoprecipitation and sequencing (ChIP-seq) protocols for high-throughput genome-wide identification of binding sites for individual transcription factors. However, we will first discuss the main experiment-based and synthetic datasets that have been available so far, as well as the benchmarking procedures that have been defined based on these data. We will then proceed by discussing how the availability of genome-wide ChIP-seq datasets may lead to more realistic and sensitive benchmarking.

3.1. *Criteria for good benchmark datasets*

As the purpose of benchmarking is to give indications on what performance can be expected in real scenarios, four basic requirements should be met:

1. Datasets should not differ too much from typical real cases.
2. Datasets should represent challenging but solvable scenarios.
3. Resulting scores should be unbiased with respect to motif discovery method.
4. Resulting scores should be reasonably robust with respect to minor changes in algorithm or parameters.

The first requirement that benchmark sets have to be representative is obvious, but due to the broad usage of motif discovery methods it is not straightforward to define precisely. The second requirement means that we should focus on those datasets where it is reasonable to assume that motif discovery may be able to find a relevant solution. For instance, as single motif discovery is typically based on overrepresentation of patterns, the dataset needs to be of a certain size in order for the binding sites to stand out from random variability. Having datasets that are reasonably large and diverse and with sufficient conservation between binding sites means that motif discovery should be possible in theory, leaving it to the methods to meet this challenge. The third requirement means that a dataset should not favor a certain set of discovery approaches. This is typically the case with synthetic benchmark sets, as discussed below. Such biases could also result from the procedures used to select and process experimental data to be used for benchmarking. The fourth requirement basically means that a benchmark suite should be sufficiently large and varied to give robust and reproducible results, and not too sensitive to nonsignificant fluctuations in output caused e.g. by the stochastic nature of some methods.

However, it is important to realize that even if the dataset itself is unbiased, it can still be used in a biased way. There are several examples of published motif discovery methods where the performance is optimal mainly on a specific dataset published together with the method, and it is tempting to believe that effects related to overtraining are involved. Most method development includes some implicit training in terms of parameter tuning and choice of background model. To avoid overtraining, it is important to do this on a separate dataset (training set) and use the benchmark set only for benchmarking (test set). If this is difficult to realize then cross validation may be an alternative, using a random subset for training and the remaining data for testing, and repeating this over several splits of the dataset.

3.2. Substring-based datasets

Most available benchmark sets are what we could call substring-based, meaning that they use substrings of DNA (e.g. promoter regions) with annotated binding sites. Sometimes the genomic location of each substring may be known, but in many cases only the DNA string itself and the binding site positions within the string will be available.

Table 2 lists a selection of recent benchmark suites, showing the number of datasets, the minimum number of sequences per dataset, as well as general benchmark type. The single motif discovery benchmarks only

Table 2. Some important benchmark datasets.

Benchmark	# seq sets	Min # seqs	Comment
Single motifs			
(Pevzner and Sze, 2000)	8	20	Synthetic
(Hu *et al.*, 2005)	62	2	*E. coli*
(Tompa *et al.*, 2005)	52	2	TRANSFAC
(Sandve *et al.*, 2007)			
Algorithm	50	5	TRANSFAC
Model	25	18	TRANSFAC
Regulatory modules			
(Krivan and Wasserman, 2001)	1	12	Liver-specific
(Wasserman and Fickett, 1998)	1	24	Muscle-specific
(Ivan *et al.*, 2008)	33	4	*Drosophila*
(Klepper *et al.*, 2008)	10	5	TRANSCompel

provide sets of sequences with shared regulatory motifs. The composite motif discovery benchmark (Klepper *et al.*, 2008) also provides binding motifs for relevant as well as randomly selected transcription factors, in order to control the level of "noise" in the benchmark.

3.2.1. *Synthetic datasets*

Synthetic datasets may be an attractive alternative for benchmarking as they allow full control over dataset properties. Given an acknowledged model of real data, it is possible to benchmark performance of different methods on synthetic data generated according to alternative parameterizations of the model. The problem with this approach is that real data are complex, varied and not accurately understood. There are almost as many suggestions on appropriate data models as there are methods. By generating datasets according to a given model, one can give an unintended advantage to methods that rely on that same model for motif discovery. This may not be appropriate, in particular since although a set of methods may be based on the same particular model, they may have varying robustness when input data do not match this assumed model. Many different synthetic datasets have been accompanying newly proposed methods, but few synthetic datasets have been used in independent benchmarking. A synthetic benchmark suite based on the *k*-mismatch model (string matching allowing up to *k* mismatches) has been proposed (Pevzner and Sze, 2000). Some benchmarks (e.g. Tompa *et al.*, 2005; Sandve *et al.*, 2007) include semi-synthetic datasets where real binding sites have been inserted into synthetic background sequences, generated by a Markov model.

3.2.2. *Single motifs*

A seminal benchmark by Tompa *et al.* (2005) represented the first broad comparison of motif discovery methods, with 13 commonly used methods being evaluated. The benchmark suite consisted of 52 organism-specific datasets (together with four negative controls) compiled from the TRANS-FAC database (Matys *et al.*, 2003). Although having organism-specific datasets has advantages, it also enforces restrictions on dataset compilation. Thus, many of the datasets used in the comparison were very small, with only 2 sequences in the smallest datasets. This influences the robustness of the benchmarking, but it also raises the question whether it is reasonable to expect any meaningful results from motif discovery on some of these datasets.

An alternative benchmark suite was proposed by Sandve *et al.* (2007). This benchmark was also based on experimental data from TRANSFAC, but made some alternative choices for compiling the datasets. First, it combined regulatory regions having instances of the same motif in related organisms, which gave considerably larger datasets. It then analyzed upper limits to motif discovery performance according to a given model, and made two groups of data depending on the ability of the chosen model to distinguish between annotated binding sites and remaining sequence in each set. This gave datasets that were possibly less realistic, due to the mixed origin of data, but more robust and more controlled with respect to expected optimal performance compared to the datasets used by Tompa *et al.* Evaluation of benchmark performance based on these datasets is available as a web-based service (http://tare.medisin.ntnu.no/).

3.2.3. Regulatory modules

Transcription factors seldom work in isolation but cooperate with other factors to regulate genes in specific ways. A collection of nearby binding sites for transcription factors that work in concert is called a *cis-regulatory module* (CRM). A gene can have several associated CRMs that regulate the gene in different contexts. Discovering such *cis*-regulatory modules computationally is the next logical step up from discovering single transcription factor binding sites, and several programs have been developed for this problem. However, the complex nature of CRMs makes these elements laborious to identify experimentally, and the availability of data on verified regulatory modules is limited.

One of the oldest sources of CRM data is a set of about fifty CRMs that regulate gene expression along the anterior–posterior axis during *Drosophila* blastoderm development. The blastoderm CRMs have been included, in addition to many other modules, in the REDfly database, a publicly available resource on regulatory elements in *Drosophila* (Halfon *et al.*, 2008). REDfly currently contains 737 CRMs associated with 249 genes. These modules were originally identified without regard to their constituent single binding sites, and although version 2.0 includes data about single sites taken from the former FlyReg database, the CRM regions in REDfly remain largely uncharacterized with respect to their actual binding sites. In fact, most of the CRMs lack any information about their constituent sites whatsoever. Ivan *et al.* (2008) used the data from REDfly to construct benchmark datasets for module discovery tools. The annotated CRMs were first grouped into separate datasets based on information about the expression pattern of the regulated genes, so that CRMs with similar tissue-specificity were grouped together. To avoid potential problems with other unknown CRMs in the vicinity of the annotated modules, each CRM was taken out of its original genomic context and planted into a synthetic background sequence. The background sequence was constructed by concatenating 1000 base pair (bp) segments randomly sampled from *Drosophila* noncoding regions, but with GC content similar to the original flanking sequence of the CRM. The resulting 33 datasets contained from 4 to 77 sequences each (average 16) with CRM lengths ranging from 83 to 2013 bp in sequences with total length of 10 times the size of corresponding CRM. These benchmark datasets only contain information on the location of the full module and no information on the binding sites for individual transcription factors.

Two other widely used module discovery benchmark datasets have been based on CRMs that drive tissue-specific regulation of genes in muscle and liver tissue respectively (Wasserman and Fickett, 1998; Krivan and Wasserman, 2001). These modules where taken from the genomes of various mammals and chicken and have an average length of about 100 bp. The muscle-specific CRMs consist of various combinations of binding sites for the five transcription factors Mef-2, Myf, Sp-1, SRF, and Tef, while the liver modules have binding sites mainly for CREB, HNF-1, 3 and 4 as well as a few other factors.

TRANSCompel is a database that specializes in composite motifs from a variety of sources, but the majority of the annotated modules come from human, mouse and rat (Matys *et al.*, 2006). TRANSCompel contains CRMs consisting of pairs and triplets of closely located binding sites where experiments have confirmed cooperative action between the transcription factors. The database is available both in a free version containing 322 modules and a commercially licensed version currently containing 428 modules. Klepper *et al.* (2008) derived 10 benchmark datasets from a subset of binding site pairs in TRANSCompel, where each dataset was based on sequences with similar CRMs containing binding sites for the same two transcription factors. These datasets provide information on both the location of the single sites and the full module, and

evaluation of benchmark performance is available as a web-based service
(http://tare.medisin.ntnu.no/).

3.3. *Genome-wide datasets*

The ultimate goal of motif discovery is to be able to work on a genome-
wide scale, in particular since it is often difficult to identify the correct
regulatory regions for datasets based on limited genomic substrings (e.g.
promoter regions). But even for substring-based approaches it does make
sense to develop datasets in a genome-wide context to make it easier to
integrate various types of additional information like evolutionary conservation and histone modification patterns into the prediction process. It
has been shown that such external data improve the performance of motif
discovery methods (Hawkins *et al.*, 2009; Whitington *et al.*, 2009).

Both ChIP-seq and ChIP-chip technologies are able to map
TF-binding on a genome-wide scale. However, ChIP-seq is rapidly becom-
ing the most popular method for experimental identification of transcription factor binding sites compared to ChIP-chip. In a ChIP-seq
experiment, the binding regions can typically be narrowed down to
100–300 bp, which is a major improvement compared to the regions
identified by ChIP-chip (often above 1000 bp). As more ChIP-seq data
become available, the relatively precise regions defined by ChIP-seq
should be ideal for benchmarking purposes, in particular as more data
and improved procedures may help us to identify regulatory complexes,
and not only single binding sites (Wallerman *et al.*, 2009).

The methodology behind a ChIP-seq experiment is as follows. First
cross-linking of DNA and proteins in living cells is used to capture DNA
bound by transcription factors *in vivo*. Sonification is used to fragment
the DNA. The average length of the resulting fragments is typically
200–300 bp. Immunoprecipitation using an antibody specific to the protein of interest is then used to select the DNA fragments bound by that
transcription factor. Reversion of the cross-link separates the transcription
factor from the DNA, and the resulting DNA fragments can be sequenced
using high-throughput sequencing techniques. Only 25–30 bp of each
fragment is sequenced, and this constitutes a tag. Tags are mapped to the
genome, and a cluster of tags around a certain genomic region indicates
a high probability that the TF is binding in this region. A typical region
where binding has occurred is shown in Fig. 3, where the peaks indicate

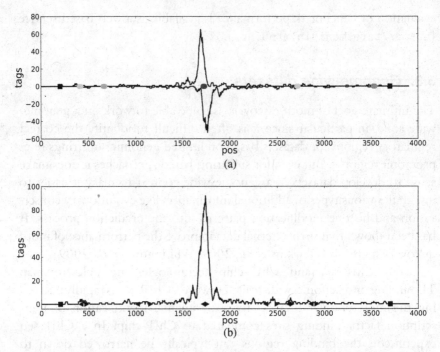

Fig. 3. Example of ChIP-seq peak data. Typical peak data from a ChIP-seq experiment. (a) Peaks as they appear on the positive (upper) and negative (lower) strand.
(b) Combined peak, where the tags on each strand are shifted towards the center. The
peak shows tag-enrichment from position 1600 to 1800. Significant binding regions
from two software tools are shown, where the squares indicate regions returned by
the program MACS (Zhang *et al.*, 2008), and the arrows indicate regions returned
by the program SISSRs (Jothi *et al.*, 2008). A region-scan by a standard PWM reveals
positions for potential binding sites, where the colors from gray to black indicate
the strength of the PWM-score. The darkest motif is a clear binding site, while
the others are ambiguous or possibly false positives. The ChIP-seq data are from a
study of NRSF (neuron-restrictive silencer factor) (Johnson *et al.*, 2007), and the
PWM used (Schoenherr *et al.*, 1996) was downloaded from the TRANSFAC database
(Matys *et al.*, 2003).

tag density. The peak-shift observed in Fig. 3 is a result of limiting the
sequencing to 25 bp in each fragment while sequencing both strands.

In order to be useful for benchmarking, the tag cluster regions have
to be identified. Several freely available software tools exist for performing this task (Jothi *et al.*, 2008; Zhang *et al.*, 2008; Tuteja *et al.*, 2009),
and typical results from two of them are shown in Fig. 3. However, some

requirements have to be fulfilled for the defined regions to be useful as a benchmark set. First, the regions have to be narrowed down to represent only the underlying peak. This is necessary to avoid too many false positive binding sites in the benchmark as described previously. Another problem in ChIP-seq data is noise, and many of the regions defined by the software do not represent relevant peaks or true binding sites. The false peaks are usually compensated for by submitting samples of nonbound DNA to the same ChIP-seq procedure, and comparing the resulting tags from the nonbound DNA with those obtained when transcription factors are bound to the DNA. This procedure reduces the number of false peaks considerably. However, additional filtering is still needed to remove artefacts that are not related to active binding sites. Finally, several well-defined peaks do not seem to include recognizable binding sites, or only weak binding sites. There may be several reasons for this: the transcription factor can, for example, bind indirectly through another factor, or participate in cooperative binding together with other factors. In the first case, there is no direct contact between the transcription factor and the DNA, but the factor will still be selected by the antibody. If the goal is to use the benchmark to predict binding sites, it should be required that the defined peak region includes at least one potential binding site. At the same time, proper evaluation and filtering of the peak regions will be required to produce a good benchmark set.

The main advantage of a ChIP-seq dataset over most existing benchmark sets is coverage. Given that a good procedure has been used for identifying significant peak regions, we should expect the identified regions to contain all active binding sites for the TF in question under that specific set of experimental conditions (cell type, metabolic state, etc.). This means that we should be able to use such datasets to test conditional prediction of binding sites on a genomic scale. This is of course different from predicting all possible binding sites in a given promoter region, independent of cell type or status, and it is also a much more challenging problem. On the other hand, ChIP-seq gives us essential data for evaluating the importance of additional information like evolutionary conservation and nucleosome positioning as input to the prediction process. ChIP-seq data also give us much more complete information on binding site status for a particular experiment, i.e. we should expect much lower numbers of unannotated active binding sites. Although this means that we are moving towards more cell-type specific datasets, as previously pointed out for predictions using chromatin modification data (Whitington *et al.*, 2009), this is probably an important and necessary step towards a more realistic setting for motif discovery.

4. Benchmarking Without a Benchmark Dataset

Most of the benchmarking approaches discussed so far are based on the assumption that we are able to define a "gold standard" reference set, i.e. a well-annotated sequence set where all binding sites are known and annotated. However, this may actually be a very rare situation. In a recent paper on assessment of TFBS prediction using phylogenetic motif models (Hawkins *et al.*, 2009), the initial conclusion was that using phylogenetic information during TFBS prediction did not improve performance for any of the methods that were tested. This was quite unexpected, as it has been generally assumed that sequence conservation across species would help to identify significant binding sites. The most likely explanation seems to be that there are more weak binding sites than expected in most sequences. Normally only strong binding sites will be known and annotated. However, more sensitive prediction methods will excel mainly in the prediction of weak sites. In standard approaches to benchmark assessment, these predictions will be incorrectly classified as false positives and the method will therefore get a low performance score. Improved reference data, e.g. from ChIP-seq analysis, may improve the situation, but we do not know yet whether all relevant weak binding sites will be populated under the experimental conditions used for a given ChIP-seq experiment. A synthetic dataset can be an alternative if we are able to give a good definition of weak binding sites relative to strong binding sites and nonbinding positions. This is normally not the case, although recent experimental data may give some relevant input to such approaches (Badis *et al.*, 2009). The alternative approach used by Hawkins *et al.* (2009) was to estimate the null distribution of motif scores, based on previous work on regular expression motifs (Kheradpour *et al.*, 2007). Then no information about the actual binding site positions is needed.

In order to estimate the null distribution of scores for a given motif, the motif was shuffled in a controlled way (mainly to maintain the general conservation profile of the motif), and 100 motif variants were tested by scanning against reference sequences, in this case from yeast. The motifs where the number of "hits" was comparable to the original motif were then clustered together with known motifs from the same organism, and the clusters containing known motifs were discarded. From the remaining clusters, 20 representative motifs were selected, and these were used to estimate the null distribution of scores for the real motif. Using this approach, they were able to show that the inclusion of phylogenetic information indeed improved prediction performance, as expected.

Although this is a very interesting approach, there are also several potential problems. It may not be possible to find enough shuffled variants for all motifs, e.g. because the motif itself has too much internal similarity, or because the performance of the shuffled variants is too different from the real motif, or because the shuffled variants are not different enough from other real motifs. In order to design adequate benchmark datasets, we have to get a better understanding of the importance of weak binding. How weak can a binding site be and still be relevant? How important are cooperative effects in weak binding sites? At least until we have more experimental data on this, these "gold standard" free approaches may actually be an important addition to traditional benchmarking.

5. Related Areas

Although the focus of this chapter is on benchmarking of TFBS prediction methods, there are several associated areas where benchmarking is essential because prediction performance may influence the quality of TFBS prediction. Methods predicting genomic features that are correlated with TF binding may for example be used as prior information in TFBS prediction, and the quality of the prior information will then have consequences for the performance of the TFBS prediction. One example is the identification of gene regulatory regions. Although TFBS prediction methods ideally should work well on genome-wide data, this is normally not the case. We can therefore improve performance if we can focus on active regulatory regions. Several methods have been developed for predicting transcription start site (TSS) and the core promoter region of a gene (e.g. Abeel *et al.*, 2008; Abeel *et al.*, 2009), and similar approaches may be used to identify regulatory regions within the gene as well as distal enhancer regions. Within a given regulatory region, it is also relevant to predict the position of nucleosomes (e.g. Segal *et al.*, 2006) in order to identify available binding sites. The quality of such approaches will both directly and indirectly influence the benchmarking discussed here, as it may affect the selection of genome regions used for prediction, the association between specific genes and their regulatory regions and the distinction between true and false binding sites.

6. Conclusion

The choice of methods for prediction of transcription factor binding sites should be based on an objective and unbiased assessment of performance. The most obvious approach to this is benchmarking against a reference dataset. However, this requires that the benchmarking itself is realistic and unbiased. In a recent study we showed (Sandve *et al.*, 2007) that the data used by Tompa *et al.* (2005) potentially could underestimate the performance of motif discovery tools because it was not possible to separate annotated binding sites from nonbinding sites with standard motif representations for several of the subsets. On the other hand, by using data where we knew that an optimal classification in principle was possible, we could also show that none of the tested methods were able to find that solution. This shows the importance of using tested and verified datasets for benchmarking in order to draw reliable conclusions.

Here we have tried to give a general overview of the current status in benchmarking of motif discovery methods. We have also tried to point out some future directions for the area, in particular with respect to genome-wide approaches and benchmarking of methods that include additional genomic information into the prediction process. A lot of work remains before we have an optimal approach for this. But without improved benchmarking tools, motif discovery may remain an unsolved problem for many years to come.

Abbreviations

ANP, ASP — average nucleotide and site performance
bp — base pairs
CC — correlation coefficient
ChIP-chip — chromatin immunoprecipitation with microarray analysis
ChIP-seq — chromatin immunoprecipitation with sequencing
CRM — *cis*-regulatory module
NRSF — neuron-restrictive silencer factor

PC — performance coefficient
PPV, NPV — positive and negative predictive value
PSSM — position specific scoring matrix
PWM — position weight matrix
Sn, Sp — sensitivity, specificity
TF — transcription factor
TFBS — transcription factor binding site
TN, TP, FN, FP — true negative, true positive, false negative, false positive
TSS — transcription start site

Acknowledgments

This work has been supported by the Norwegian Research Council through the FUGE programme.

References

Abeel T, Saeys Y, Bonnet E, Rouze P, Van de Peer Y. (2008) Generic eukaryotic core promoter prediction using structural features of DNA. *Genome Res* **18**: 310–323.

Abeel T, Van de Peer Y, Saeys Y. (2009) Toward a gold standard for promoter prediction evaluation. *Bioinformatics* **25**: i313–320.

Ao W, Gaudet J, Kent WJ, Muttumu S, Mango SE. (2004) Environmentally induced foregut remodeling by PHA-4/FoxA and DAF-12/NHR. *Science* **305**: 1743–1746.

Badis G, Berger MF, Philippakis AA *et al.* (2009) Diversity and complexity in DNA recognition by transcription factors. *Science* **324**: 1720–1723.

Bailey TL, Williams N, Misleh C, Li WW. (2006) MEME: Discovering and analyzing DNA and protein sequence motifs. *Nucleic Acids Res* **34**: W369–373.

Bolstad KH. (2009) Comparison of strategies for benchmarking of motif discovery methods. Master's thesis. Department of Biology, Norwegian University of Science and Technology, Trondheim.

Halfon MS, Gallo SM, Bergman CM. (2008) REDfly 2.0: An integrated database of cis-regulatory modules and transcription factor binding sites in Drosophila. *Nucleic Acids Res* **36**: D594–598.

Hawkins J, Grant C, Noble WS, Bailey TL. (2009) Assessing phylogenetic motif models for predicting transcription factor binding sites. *Bioinformatics* **25**: i339–347.

Hu J, Li B, Kihara D. (2005) Limitations and potentials of current motif discovery algorithms. *Nucleic Acids Res* **33**: 4899–4913.

Ivan A, Halfon MS, Sinha S. (2008) Computational discovery of cis-regulatory modules in Drosophila without prior knowledge of motifs. *Genome Biol* **9**: R22.

Johnson DS, Mortazavi A, Myers RM, Wold B. (2007) Genome-wide mapping of *in vivo* protein-DNA interactions. *Science* **316**: 1497–1502.

Jothi R, Cuddapah S, Barski A, Cui K, Zhao K. (2008) Genome-wide identification of *in vivo* protein-DNA binding sites from ChIP-Seq data. *Nucleic Acids Res* **36**: 5221–5231.

Kheradpour P, Stark A, Roy S, Kellis M. (2007) Reliable prediction of regulator targets using 12 Drosophila genomes. *Genome Res* **17**: 1919–1931.

Klepper K, Sandve GK, Abul O, Johansen J, Drablos F. (2008) Assessment of composite motif discovery methods. *BMC Bioinformatics* **9**: 123.

Krivan W, Wasserman WW. (2001). A predictive model for regulatory sequences directing liver-specific transcription. *Genome Res* **11**: 1559–1566.

Mahony S, Benos PV. (2007) STAMP: A web tool for exploring DNA-binding motif similarities. *Nucleic Acids Res* **35**: W253–258.

Matys V, Fricke E, Geffers R *et al.* (2003) TRANSFAC: Transcriptional regulation, from patterns to profiles. *Nucleic Acids Res* **31**: 374–378.

Matys V, Kel-Margoulis OV, Fricke E *et al.* (2006) TRANSFAC and its module TRANSCompel: Transcriptional gene regulation in eukaryotes. *Nucleic Acids Res* **34**: D108–110.

Pavesi G, Mereghetti P, Mauri G, Pesole G. (2004) Weeder Web: Discovery of transcription factor binding sites in a set of sequences from co-regulated genes. *Nucleic Acids Res* **32**: W199–203.

Pevzner PA, Sze SH. (2000) Combinatorial approaches to finding subtle signals in DNA sequences. *Proc Int Conf Intell Syst Mol Biol* **8**: 269–278.

Roth FP, Hughes JD, Estep PW, Church GM. (1998) Finding DNA regulatory motifs within unaligned noncoding sequences clustered by whole-genome mRNA quantitation. *Nat Biotechnol* **16**: 939–945.

Sandelin A, Alkema W, Engstrom P, Wasserman WW, Lenhard B. (2004). JASPAR: An open-access database for eukaryotic transcription factor binding profiles. *Nucleic Acids Res* **32**: D91–94.

Sandve GK, Abul O, Walseng V, Drablos F. (2007) Improved benchmarks for computational motif discovery. *BMC Bioinformatics* **8**: 193.

Sandve GK, Drablos F. (2006) A survey of motif discovery methods in an integrated framework. *Biol Direct* **1**: 11.

Schoenherr CJ, Paquette AJ, Anderson DJ. (1996) Identification of potential target genes for the neuron-restrictive silencer factor. *Proc Natl Acad Sci USA* **93**: 9881–9886.

Segal E, Fondufe-Mittendorf Y, Chen L *et al.* (2006) A genomic code for nucleosome positioning. *Nature* **442**: 772–778.

Thijs G, Lescot M, Marchal K *et al.* (2001) A higher-order background model improves the detection of promoter regulatory elements by Gibbs sampling. *Bioinformatics* **17**: 1113–1122.

Tompa M, Li N, Bailey TL *et al.* (2005) Assessing computational tools for the discovery of transcription factor binding sites. *Nat Biotechnol* **23**: 137–144.

Tuteja G, White P, Schug J, Kaestner KH. (2009) Extracting transcription factor targets from ChIP-Seq data. *Nucleic Acids Res* **37**: e113.

Wallerman O, Motallebipour M, Enroth S *et al.* (2009) Molecular interactions between HNF4a, FOXA2 and GABP identified at regulatory DNA elements through ChIP-sequencing. *Nucleic Acids Res* **37**(22): 7498–7508.

Wasserman WW, Fickett JW. (1998) Identification of regulatory regions which confer muscle-specific gene expression. *J Mol Biol* **278**: 167–181.

Wasserman WW, Sandelin A. (2004) Applied bioinformatics for the identification of regulatory elements. *Nat Rev Genet* **5**: 276–287.

Whitington T, Perkins AC, Bailey TL. (2009). High-throughput chromatin information enables accurate tissue-specific prediction of transcription factor binding sites. *Nucleic Acids Res* **37**: 14–25.

Zhang Y, Liu T, Meyer CA *et al.* (2008) Model-based analysis of ChIP-Seq (MACS). *Genome Biol* **9**: R137.

Chapter 7

Encyclopedias of DNA Elements
for Plant Genomes

Jens Lichtenberg*,††, Alper Yilmaz†, Kyle Kurz*,
Xiaoyu Liang*, Chase Nelson‡, Thomas Bitterman§,
Eric Stockinger¶, Erich Grotewold†
and Lonnie R. Welch*,||,**,‡‡

This chapter focuses on the enhancement of existing plant regulatory databases by adding DNA word encyclopedias. The authors present DNA word encyclopedias for small sets of co-regulated sequences (C-repeat binding factor genes in wheat) and for intergenic regions of an entire genome (*Arabidopsis thaliana*). It is also shown how the resulting DNA word encyclopedia for *Arabidopsis* is incorporated into an existing repository, the Arabidopsis Gene Regulatory Information Server (AGRIS). This provides a model for how DNA word encyclopedias can be incorporated into organism-specific regulatory databases.

1. Introduction

The *cis*-regulatory code corresponds to the set of hardwired DNA instructions necessary for the expression of all genes. The *cis*-regulatory code can

*Bioinformatics Laboratory, School of Electrical Engineering and Computer Science, Ohio University, Athens, OH, USA
†Plant Cellular and Molecular Biology Department, The Ohio State University, Columbus, OH, USA
‡Department of Biology, Oberlin College, Oberlin, OH, USA
§Cyberinfrastructure Group, Ohio Supercomputer Center, Columbus, OH, USA
¶Department of Horticulture and Crop Science, The Ohio State University, Ohio Agricultural Research and Development Center, Wooster, OH, USA
||Biomedical Engineering Program, Ohio University, Athens, OH, USA
**Molecular and Cellular Biology Program, Ohio University, Athens, OH, USA
††lichtenj@ohio.edu
‡‡welch@ohio.edu

be conceptualized as being formed by short DNA-sequence words and motifs — the *cis*-regulatory elements (CREs) — which can mediate the recruitment of *trans*-acting regulatory proteins — the transcription factors (TFs) — to gene regulatory regions. CREs can also affect gene expression at other levels, for example by serving as docking sites for small interfering RNAs (siRNAs), or by affecting chromatin structure.

Cis-regulatory elements are DNA words that are often shared among highly co-regulated genes (Liang, *et al.*, 2010) or among functionally similar genes (Lichtenberg *et al.*, 2009a). Based on this idea, DNA word encyclopedias can be created for small sets of related promoter regions. In this paper, we present a DNA encyclopedia for the C-repeat binding factor genes in *Triticeae*. Additionally, the authors establish a DNA word encyclopedia for *Arabidopsis thaliana*. Together with existing functional information on specific *cis*-regulatory elements, the word encyclopedia for *Arabidopsis* is incorporated into the Arabidopsis Gene Regulatory Information Server (AGRIS) (Davuluri *et al.*, 2003). This not only provides a valuable resource for the reference organism *Arabidopsis thaliana*, but also serves as a model for other organism-specific regulatory databases.

2. C-repeat Binding Factor Genes in *Triticeae*

The capacity to survive freezing temperatures is a trait essential for the winter-grown cereal crop plants wheat and barley. Central to gene regulatory network pathways affecting freezing tolerance are the CBF (C-Repeat Binding Factor) transcription factors (Thomashow, 2001). More than 20 *CBF* genes distributed across multiple phylogenic clade occur in the diploid genome of barley (Skinner *et al.*, 2005). *CBFs* in the CBFIIIa, CBFIIId, CBFIVa, CBFIVc, and CBFIVd clades are induced by low temperature (Badawi *et al.*, 2007; Stockinger *et al.*, 2007). Induction is affected by a circadian clock and is repressed by the developmental state (Stockinger, *et al.*, 2007). *CBFs* in the CBFIIIc clade are non-responsive to low temperature (Badawi *et al.*, 2007; Stockinger *et al.*, 2007). Identifying the DNA elements controlling *CBF* expression is an essential step towards understanding the molecular processes regulating *CBF* expression, and in turn will enable strategies to improve the freezing tolerance of wheat and barley.

CBF promoters were divided into three groups based on phylogeny and expression patterns (Table 1). Using 1250 bp upstream of the ATG

Table 1. Identified motifs for CBF gene in *Triticeae*.

Word (R-score)[a]	CBFIIIa and CBFIIId (A)		CBFIIIc (B)		CBFIVa, CBFIVc, and CBFIVd (C)		All cold responsive		All promoters	
	7		7		8		15		22	
	N[b]	O	N	O	N	O	N	O	N	O
CGCGGT(0.3677)[a]	7[d]	24	7	12	6	9	13	33	20	45
ACCGCGT	6	7	0	0	0	0	6	7	6	7
ACGCGTC	6	11	2	2	1	1	7	12	9	14
AGCGCGTTCATACAC	0	0	5	5	0	0	0	0	5	5
CACCGCGT	4	5	0	0	0	0	4	5	4	5
CGCGTC	6	15	4	4	2	4	8	19	12	23
ACGCG(0.3523)	7[d]	25	4	6	6	15	13	40	17	46
AACGCGG	6	7	0	0	1	1	7	8	7	8
ACGCGTC	6	11	2	2	1	1	7	12	9	14
CACTC (0.3642)	7[d]	24	7	33	6	11	13	35	20	68
ACACTC	6	15	7	17	2	2	8	17	15	34
AGCACTCTG	0	0	5	5	0	0	0	0	5	5

Number of sequences in dataset

(*Continued*)

Table 1. (*Continued*)

Word (R-score)[a]	CBFIIIa and CBFIIId (A) 7		CBFIIIc (B) 7		CBFIVa, CBFIVc, and CBFIVd (C) 8		All cold responsive 15		All promoters 22	
	N[b]	O	N	O	N	O	N	O	N	O
CACACTC	5	6	7	12	1	1	6	7	13	19
CTCAA (0.2597)	7	21	7[d]	26	7	12	14[d]	33	21	59
CTCAAGC	3	4	7	13	1	2	4	6	11	19
TAAGCTCAAGCA	0	0	5	5	0	0	0	0	5	5
TCAAGCTCAA	5	6	5	5	0	0	5	6	5	5
CTCCA (0.3512)	7	20	7	17	8[d]	29	15[d]	49	22[d]	66
ACTCCA	6	12	5	6	6	8	12	20	17	26
CTCCAC	3	3	4	6	7	17	10	20	14	26
GCTCCAC	0	0	0	0	5	7	5	7	5	7
CTTGT (0.3409)	7	20	5	10	8[d]	24	15[d]	44	20	54
GCGTCACTTGTC	3	5	0	0	0	0	3	5	3	5
CACCG (0.3272)	6	21	6	13	8[d]	17	14[d]	38	20	51
CACGGC	5	9	3	5	8	8	13	17	16	22

Number of sequences in dataset

(*Continued*)

Table 1. (*Continued*)

Word (R-score)[a]	CBFIIIa and CBFIIId (A) 7		CBFIIIc (B) 7		CBFIVa, CBFIVc, and CBFIVd (C) 8		All cold responsive 15		All promoters 22	
	N[b]	O	N	O	N	O	N	O	N	O
CACCGCGT	4	5	0	0	0	0	4	5	4	5
CCACCG	5	8	1	1	6	11	11	19	12	20
GCTTG(0.2763)	7	15	3	4	7	19	14[d]	34	17	38
TGCTTG	6	6	1	1	6	9	12	15	13	16
CAAGC(0.4219)	7	21	7	35	7	14	14	35	21[d]	70
AGCACCAAGC	0	0	4	5	0	0	0	0	4	5
CAAGCT	5	7	7	16	5	6	10	13	17	29
CTCAAGC	3	4	7	13	1	2	4	6	11	19
TAAGCTCAAGCA	0	0	5	5	0	0	0	0	5	5
TCAAGCTCAA	0	0	5	5	0	0	0	0	5	5

Number of sequences in dataset

[a] The first five 5-bp motifs were predicted with *maximum confidence* because the word (1) scored high, (2) appeared frequently within a single sequence, and (3) appeared in multiple sequences; the latter four motifs met only two of these three criteria and were predicted with *high confidence*.

[b] Motif frequency: N = Number of promoters possessing motif, O = Number of occurrences in promoters.

[c] The 5-bp core motifs are in the top row of each column section. They are followed by the larger motifs within which the 5-bp motif occurred.

[d] Dataset within which motif score is maximal or high.

translation initiation codon, we searched for motifs that were at least 5-bp in length, invariant at a minimum of five positions, and were present in a minimum number of sequences. Nine of the identified 5-bp motifs were predicted to be biologically relevant (Table 1), each motif embedded in larger motifs that were also identified during the analysis. The larger motifs differentiated the three *CBF* groups and were usually present in only one group. For example, the 5-bp motif CGCGT occurred within the larger motifs AC**CGCGT**, A**CGCGT**C, and **CGCGT**C present in the low-temperature responsive CBFIIIa and CBFIIId *CBF* promoters (Table 1). The 5-bp motif was present in all seven low-temperature responsive CBFIIIa and CBFIIId promoters, appearing 24 times in total in this group (Table 1). The 5-bp core motif was also present in all of the low-temperature non-responsive CBFIIIc promoters, appearing 12 times in total, and in six of the eight CBFIVa, CBFIVc, and CBFIVd promoters. However, the 5-bp core motif in the low-temperature non-responsive CBFIIIc promoters harbored nucleotides 5′ (AG) and 3′ (TCATACAC) to the 5-bp core motif resulting in a much larger motif, AG**CGCGT**TCATACAC, that was exclusive to the CBFIIIc promoters (Table 1).

Querying the 5-bp identified motifs against the PLACE (plant *cis*-acting regulatory DNA elements) database (Higo *et al.*, 1999) indicated that the CGCGT motif and three others occurred as part of the consensus motif in known transcription factor binding sites. Two of the 5-bp cores were reverse compliments differing in their flanking nucleotides; five were novel. The CGCGT motif plays a role in Ca^{2+}-responsiveness and is a Calmodulin-Binding Transcription Activator (CAMTA) binding site (Wang *et al.*, 2002; Kaplan *et al.*, 2006; Doherty *et al.*, 2009). CAMTA proteins were recently identified as key regulators of *Arabidopsis thaliana CBF* expression and effectors of freezing tolerance (Doherty *et al.*, 2009). The ACGCG, CACCG, and CACTC motifs are implicated in ABA-responsive and embryo-specification elements (Kim *et al.*, 1997). The CACTC motif is also implicated as an initiator element in TATA-less promoters (Nakamura *et al.*, 2002).

Additional patterns were also detected. The larger motifs, unique and structurally conserved across the low temperature non-responsive CBFI-IIc promoters, with the exception of the AG**CGCGT**TCATACAC motif, occurred either within the 5′ UTR, or flanked the TATA box (not shown). Motifs in the low-temperature responsive *CBF*s were randomly distributed (not shown). Additionally, most CBFIII subgroup *CBF* promoters (low-temperature–responsive and low-temperature–nonresponsive) harbored a positionally-correct TATA box motif having the configuration TATAAA, whereas most of the CBFIV subgroup CBFs had the TATA box configuration of TATATA (not shown).

The identification of 5-bp core motifs common to the different *CBF* promoter groups suggests the trans-acting factors interacting with the different promoters may be shared while the presence of group-specific base pairs flanking the core motifs suggests an additional level of regulatory control. These identified motifs are currently being used in functional screens to identify factors binding them.

3. Analysis of the Non-coding Segments in *Arabidopsis thaliana*

The word landscape of noncoding segments of *Arabidopsis thaliana* (based on TAIR version 9) was generated, updating the authors' previously published *Arabidopsis* word landscape of 8-mers (Lichtenberg *et al.*, 2009b) to include word lengths of 5–15 bp. The resulting word landscape has been used to populate the AGRIS database.

Based on the approach outlined in (Lichtenberg *et al.*, 2009b; Welch *et al.*, 2009), the top five words are extracted and presented for each segment (Table 2), while the complete results are made available in the database. The results are sorted based on the $S*\ln(S/E\_s)$ score and filtered for words with a *p*-value of 0.05 or below to ensure its statistical significance. The $S*\ln(S/E\_s)$ score provides a measure of coverage for each word by determining how over-represented the word is with regard to the number of sequences containing it. Additionally, the occurrence of each word is captured in the O/E and $O*\ln(O/E)$ scores.

An analysis of the top five statistically significant words for the different segments reveals that six of them are shared across different segments, with two of them in two segments (AATATT, GAAAAAG), two of them in three segments (CAAAAAC, GTTTTG) and also two of them in four segments (CAAAAC, GTTTTTG). While a strong overlap between the proximal and distal promoters can be determined based on the top five significant words, the remaining segments show little to no overlap, with

Table 2. Top five words for each of the noncoding segments and the genome of *Arabidopsis thaliana*.

Segment	Word	S	E_s	O	E	O/E	O*ln(O/E)	S*ln(S/E_s)	RevComp	Pval
3' UTRs	CTTTTG	4755	4160.29	5682	4989.47	1.1388	738.514	635.328	CAAAAG	2.22E-16
	GTTTTG	5147	4589.77	6377	5602.8	1.13818	825.38	589.765	CAAAAC	3.33E-16
	GTTTTTG	2129	1614.09	2322	1760.4	1.31902	642.937	589.477	CAAAAAC	0
	ATTTTG	5320	4781.61	6432	5883.9	1.09315	572.875	567.617	CAAAAT	9.67E-13
	ATTTTA	4395	3863.93	5257	4578.6	1.14817	726.344	565.998	TAAAAT	1.11E-16
5' UTRs	CTCTTT	3608	2979.93	4622	3623.72	1.27549	1124.66	690.041	AAAGAG	2.22E-16
	ACCCTA	1294	763.683	1381	824.08	1.67581	713.004	682.379	TAGGGT	4.44E-16
	TTCTCC	2910	2302.65	3391	2696.69	1.25747	776.874	681.214	GGAGAA	1.55E-15
	CAAAAAC	1083	683.968	1132	740.7971	1.52808	479.985	497.725	GTTTTTG	0
	CAAAAC	2316	1869.71	2643	2138.81	1.23573	559.426	495.753	GTTTTG	4.44E-16
Introns	GTAAGT	17853	13144.5	18484	14919.8	1.23889	3959.58	5465.98	ACTTAC	7.77E-16
	TTGCAG	21003	16440.4	21843	19090.2	1.1442	2942.33	5144.19	CTGCAA	0
	TTTCAG	21722	17297.3	23673	20204.2	1.17169	3750.84	4947.76	CTGAAA	3.33E-16
	GTTTTTG	13570	9789.74	14939	10929.2	1.36689	4669.04	4430.96	CAAAAAC	7.77E-16
	GTTTTG	31987	27849.5	40244	34987.5	1.15024	5632.94	4430.71	CAAAAC	1.78E-15
Core promoters	TATAAA	5284	4351.35	5675	4987.17	1.13792	733.22	1026.13	TTTATA	0
	TAAAAT	3421	2726.78	3822	3023.08	1.26427	896.25	775.923	ATTTTA	1.11E-16
	CAAAAC	2740	2089.76	2984	2287.99	1.3042	792.52	742.287	GTTTTG	1.11E-16
	TATAAAT	2249	1634.88	2298	1793.17	1.28153	570.027	717.241	ATTTATA	2.22E-16
	AATATT	2664	2057.7	2869	2251.49	1.27427	695.365	687.946	AATATT	8.88E-16

(Continued)

Table 2. (*Continued*)

Segment	Word	S	E_s	O	E	O/E	O*ln(O/E)	S*ln(S/E_s)	RevComp	Pval
Proximal promoters	CAAAAC	13854	12338.4	20348	16534.4	1.23065	4223.06	1605.13	GTTTTG	1.67E-15
	GTTTTTG	5740	4378.95	6608	4806.62	1.37477	2103.23	1553.54	CAAAAAC	1.67E-15
	CAAAAAC	6136	4769.88	7314	5279.79	1.38528	2383.67	1545.36	GTTTTTG	0
	GTTTTG	13057	11640.8	18875	15279.4	1.23533	3988.94	1499.04	CAAAAC	9.99E-16
	GAAAAAG	5538	4310.83	6384	4725.01	1.35111	1921.1	1387.28	CTTTTTC	0
Distal promoters	GTTTTTG	10688	8475.7	14445	10188.4	1.41779	5042.7	2478.75	CAAAAAC	8.88E-16
	CTTTTTC	9891	7848.09	12682	9288.69	1.36532	3948.99	2288.33	GAAAAAG	3.33E-16
	CAAAAAC	10259	8217.71	13765	9814.97	1.40245	4655.6	2276.1	GTTTTTG	0
	GAAAAAG	9761	7791.02	12591	9208.33	1.36735	3939.39	2200.35	CTTTTTC	4.44E-16
	GTTCTTG	6425	4930.29	7728	5456.47	1.4163	2689.71	1701.34	CAAGAAC	2.22E-16
Genome	TTTTTT	5	5	372939	319693	1.16655	57453	AAAAAA	Yes	
	TTTTTTT	5	5	209185	173079	1.20861	39634	AAAAAAA	Yes	
	TATATA	5	5	179209	145199	1.23423	37714	TATATA	Yes	
	CAAAAC	5	5	89769	74581.8	1.20363	16638	GTTTTG	Yes	
	AATATT	5	5	118527	105726	1.12107	13546	AATATT	Yes	

Table 3. Overlap between the top five significant words and the segments.

Word	3′ UTR	5′ UTR	Intron	Core promoter	Proximal promoter	Distal Promoter	Genome
AATATT				X			X
GAAAAAG					X	X	
CAAAAAC					X	X	
GTTTTG		X			X		
CAAAAC	X						X
GTTTTTG					X	X	

the 3′ UTR exhibiting a signature that is completely independent from the other segments (Table 3).

An analysis of the shared words against the AGRIS transcription factor binding site repository revealed that two of the words are known functional elements. GTTTTG and its reverse complement, CAAAAC, map the PII promoter motif with a binding site represented by TTGGTTTTGAT-CAAAACCAA, which is palindromic. A comparison of the words against the TRANSFAC database revealed a similarity between GTTTTG and the ELF-1 binding site (M00110), as well as a match between AATATT and the FOXJ2 binding site (M00423). Expanding the analysis to incorporate all top five words of the different fragments reveals additional matches between TATATA and CF2-II (M00013), TATAAA and the TATA box (M00216, M00252, M00320, M00471) as well as the XFD-2 binding site (M00268), CTCTTT and SOX (M01014), and ACCCTA and ZAP1 (M00754). Besides the strong similarities of some words with known *cis*-regulatory elements, other words are not part of current regulatory element databases. While it is possible that these words themselves are not functional, a recent study has shown that they could be part of larger nonrandom sequence patterns, called pyknons (Feng *et al.*, 2009).

Determining the positional bias of a putative regulatory element can lead to additional evidence that the word is truly of regulatory nature by exhibiting similar positional characteristics (as does the TATA box). As shown in Table 4, a strong positional bias is observed for the interesting word extracted from the core promoters at position 70. Since core promoters have been arbitrarily assigned to 100 base pairs in length, position 70 within the extracted sequence relates to a base pair at position 30 upstream of the transcription start site, a position that has been confirmed for the

Table 4. Positional distribution for interesting words discovered in the noncoding segments.

Segment	Word	Positional bias

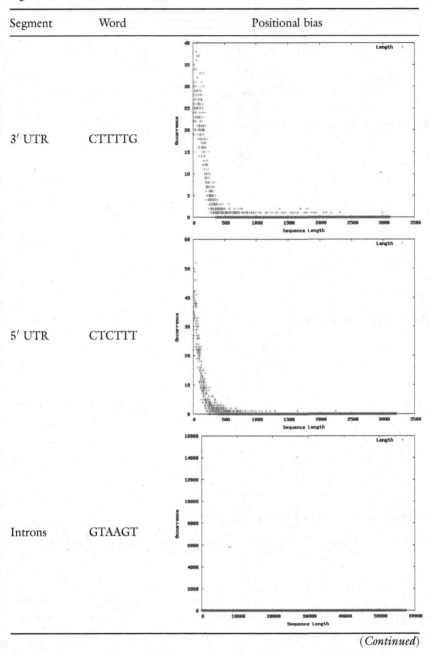

3′ UTR	CTTTTG	
5′ UTR	CTCTTT	
Introns	GTAAGT	

(Continued)

Table 4. (*Continued*)

Segment	Word	Positional bias
Core promoters	TATAAA	
Proximal promoters	CAAAAC	
Distal promoters	GTTTTG	

TATA-box in *Arabidopsis thaliana*. A spike similar to the one observed for the word embedded in the core promoters can be detected for those in the 3′ and 5′ UTRs, providing evidence for a possible regulatory behavior associated with the sites. For introns, the word GTAAGT shows a very strong peak at the start of the intron sequences, but very few occurrences afterwards, which, together with the lack of known regulatory function associated with the word, would lead to the prediction that this word could be used as an intronic recognition site in the splicing process. Finally, the words embedded in the proximal and distal promoter regions show no apparent positional bias. This, however, does not eliminate them as potential regulatory elements, since it is possible that the transcription factor binding sites are position independent.

4. Enhancement of the Arabidopisis Gene Regulatory Information Server (AGRIS)

The results of the analysis of the *Arabidopsis thaliana* genome were incorporated into the Arabidopsis Gene Regulatory Information Server for use in plant regulatory genomics research. Words of length 5–15 have been counted and analyzed, and are available for query through the AGRIS webpage http://arabidopsis.med.ohio-state.edu/.

In addition to AGRIS, there are several resources which provide details about putative or confirmed *cis*-elements in the *Arabidopsis* genome. PLACE (Higo *et al.*, 1999) lists known *cis*-regulatory elements by curating the publications. AthaMap (Steffens *et al.*, 2004) utilizes known binding site motifs and performs matrix-based and pattern-based screenings in the genome to identify potential transcription factor binding sites. These websites rely on known motifs. On the other hand, ATTED-II (Obayashi *et al.*, 2009) provides a list of heptamers based on correlation between expression and a defined group of genes, and positional information, identified by searching oligomers that are correlated with gene expression. They exhibit only 304 heptamers out of all possible heptamers and their study examined only 200 bp upstream of genes. The integration of regulatory encyclopedia data into AGRIS makes it a unique tool, distinct from other *cis*-regulatory element databases. After the integration, AGRIS contains data on all possible 5- to 15-mers analyzed in several noncoding regions.

When a user requests information about a specific word or putative regulatory element, the new AGRIS interface provides a table of statistics about that word's presence in the various segments of the *Arabidopsis* genome (Fig. 1), as well as interactive graphs of the locations where the word occurs in each segment (Fig. 1(c)). By clicking the "sequence hits" number from the column "Unique sequence occurrences" (Fig. 1(b)), AGRIS provides detailed location information of the queried word (Fig. 2). Besides including the locations of the selected word in the chosen segment, the locations of the word in the genome and the orientation are included as well. Furthermore, sequences that contained the query word are displayed with the chosen word shown in red. An additional query searches for matches between the selected word and all experimentally verified motifs stored in AGRIS, and any matches are reported to the user along with the original publication (Fig. 1(a)). If a word does not occur in a segment, the table will reflect this and an image will replace the graph for the segment where the word is missing to notify the user that no location infomation is available. This allows AGRIS to report missing words without storing them explicitly in the database. Graphs are available for the 3′ UTR, 5′ UTR, intron, core promoter, proximal promoter, and distal promoter segments.

Additionally, researchers may submit a sequence, or sequences, and see a list of words in their sequence(s), a comparison of how those words are distributed in a selected segment, and a score indicating if a word is over-represented or under-represented in the input compared to its representation in the chosen segment (Fig. 3). In the event that the words list is longer than 50, the top 50 words are displayed in the table, and sorted in descending order according to the score. In addition, a download option is provided for downloading either the original result files or the comparison result between selected segments and a user-defined query. Each word found in the input sequence(s) is queried against known binding sites and matches are marked with a double-asterisk (**). The user may click on any word found to submit a word query as described above.

To create location histograms for word queries, the javascript library Flot was used. Available at http://code.google.com/p/flot/, Flot uses jQuery plugins to create interactive plots of word locations throughout the various segments of the *Arabidopsis thaliana* genome, while maintaining the responsiveness and simplicity of the AGRIS query webpage. Each figure may contain multiple plots, and multiple figures may be created for

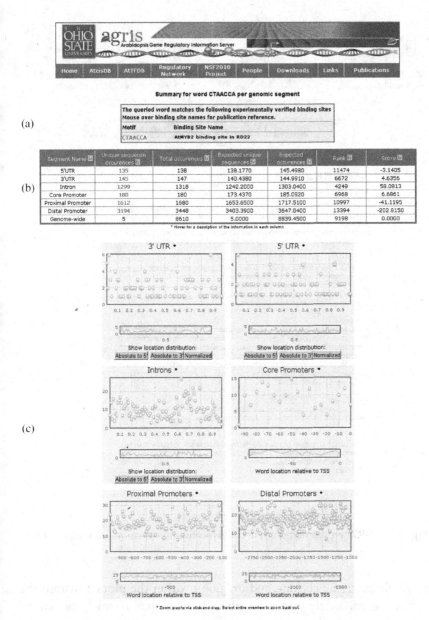

Fig. 1. AGRIS representation of a word query. (a) Representation of experimentally verified motifs that are matched with the query word. (b) Statistics of each segment for a single word. (c) Interactive location graphs in each segment for a single word.

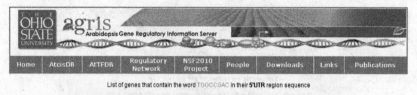

List of genes that contain the word TGGCCGAC in their **5'UTR** region sequence

GeneID	Word Location	Genomic Location *	Orientation of the Gene *
AT1G59560.1	90	21881689	F
AT3G09550.1	43	2931908	F
AT3G28900.1	8	10904443	R
AT5G01890.1	416	344704	R
AT5G47200.1	59	19166961	F
AT5G67490.1	235	26934710	R

Sequences

```
>AT1G59560.1
AATTAAAAAATTCATCCCTATTCGACTTCACTTTCTCTCTCTGCCTGAAAATTTTGATCGGTTTCTGGTTTCTCG
GCGAAAGAAGTAAAGTGGCCGACAAATAGTGATCTCGTGAAGGAATCAGTGTGTACAGTGGAAGAAG

>AT3G09550.1
ATAATCAGATCGCAAACCCTTGCCGTGACACTGAAGGCACCGGTGGCCGACGGCTTATTCATCTTTCTCCAATCG
GAGAGTCGTCGTTTTTTACAATATCATGATCCCGCTTTGAAATTTCTCTGTTTTCTAATCTCTCAAA

>AT3G28900.1
ATCCCCTTTGGCCGACTCTTACTGACCTCAGAGCTCAGCAAAA

>AT5G01890.1
GAAAAAGTATTAAAGTAGTAATAAAAAGCGGGGAGGAGGAGAAGATTGTCTCCTAGAGCGACTTGAATCCTTGAGC
TTTCAAACTCATCTCCCGGAGCTTCCCTTTTCTCTTTACTCTCTCTCTCTCTCTCTCTCTCTCACTTTTCTCT
GTTTTTTTTCTTGGTGTCTGTCTTAATTCCACCATTAATGCTCTCTACCTTCCCTTTTACTCTTCCCCTGCTTTT
GTGTTTTCTCTTTCTCTTTTCTAAACCCTTTCAAACTCTTCTTCCTCTTCTTTTCTTTTTCTCCAATAACAGTTTC
TTGTGCTCATGTTAAAACTGCAGCTCTTTTCCGGTAAAAAGAAAAGCTTTTTGGTCAGTTTAGTTATCCTCAAGC
ATCAAAACCTTACTTGAGCTCTGCATCATCATCCATGGCCATGGCCGACTAGCCTTCTTTCTTCGCTTTTTACCT
CCTTAATATCTTTAAAAGCGAGAGGAA

>AT5G47200.1
AATAAGCCAAAGCCACATAGAAGAAAAAAAAAAGAACATTCACGTCTCTCTCGTTTTTTTGGCCGACGACGATCG
CTGAATTGACTGCCGGAGATTCCTTTAATCGTCAGATTCTCGTTGAGGGATA

>AT5G67490.1
CCCTCTAATTCTCTCCTATTTGTTTTATAAACATGTTAGGCCTAGCTCCTGATGGGCTTGTTAAACATAATTATA
CTGGGCCAAATTTACTCTCTAATTCTCTTAATTTATTTATAAACATGTAAATGATTGGGCCTAGATACTACTGGG
CTTATTATCCAATTATGGCCCAACCTAAATGACTATCGAACCCTATACTATCACAAACCGTGAAGGTCGGATCTC
TTTCTTGATTTGGCCGACACATCTGATCTAACA
```

Fig. 2. AGRIS representation of detailed location information for a single word.

any single page. By clicking and dragging within the plot, or within the overview immediately under the plot, the user may zoom in or out on specific sections of the graph (Fig. 4).

In order to accurately depict location information, several graphs are created for each segment to show where the word occurred in that segment. Six different graphs are provided each time a user performs a search.

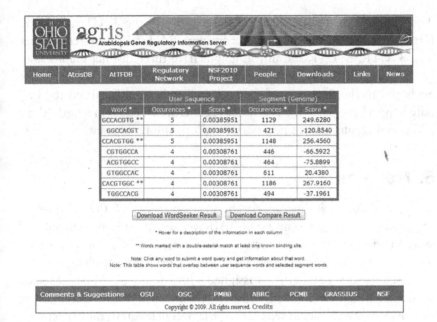

Fig. 3. Mapping of words form a user-provided sequence against information stored in AGRIS.

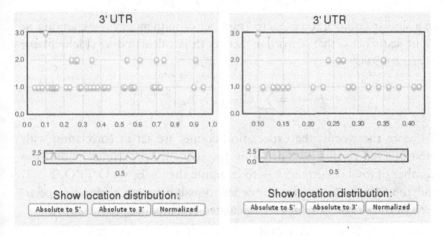

Fig. 4. Illustration of the zooming functionality for word location distributions in the enhanced AGRIS interface.

The locations of the specified word are shown in the 3′ UTR, 5′ UTR, intron, core promoter, proximal promoter and distal promoter regions. The 3′ UTR, 5′ UTR and intron graphs have three distinct graphs associated with each: one anchored to the 5′ region of the segment, one anchored to the 3′ region of the segment, and one normalized to elminate length bias. Since the promoter regions are all equal in length, these additional graphs are not necessary for the promoter regions.

5. Methods

Using a sliding window approach, the sequences of the *Arabidopsis thaliana* genome (genome build TAIR version 9) are enumerated and stored in a radix trie data structure (Morrison, 1968), allowing for scalable (in regard to memory and time complexity) word searching.

Each word w of length v is assigned a probability p_w, using a Markov model (Robin *et al.*, 2005; Ewens and Grant, 2001) of maximum order ($order = v - 2$). In a set of sequence of the overall length l, the expectation $E(w)$, for a word w of length v, of the total number of occurrences is calculated using (Lichtenberg *et al.*, 2009a):

$$E(w) = (l - v + 1) * p_w$$

In a similar fashion, the expected number of sequences $E_S(w)$, among m input sequences, that a word w occurs in is calculated as (Lichtenberg *et al.*, 2009a):

$$E_s(w) = \sum_{k=1}^{m} (1 - (1 - p_{w_i})^{l_k - v + 1})$$

To score the words, the expectation counts are set in correlation with the observed counts — S denoting the sequence occurrence and O the number of total occurrences — to compute the O/E(w), O*ln(O/E(w)) and the S*ln(S/E$_S$) scores. In order to assign statistical significance to each of the words, p-values are calculated (Lichtenberg *et al.*, 2009a):

$$p(w) = 1 - \sum_{k=1}^{m} \sum_{i=a}^{l_k - v + 1} \binom{l_k - v + 1}{i} p_w^i (1 - p)^{l_k - v + 1 - i}$$

Besides the statistical significance and over-representation, each word is looked up in the TRANSFAC database (Wingender *et al.*, 2000) and the AGRIS database (Davuluri *et al.*, 2003) of known transcription factor binding sites.

6. Conclusion

An approach for the compilation of DNA word encyclopedias was applied to several plant genomes. Additionally, it was shown how the resulting encyclopedias can be integrated with existing genomic repositories and that the enhanced databases provide valuable insights to scientists who study regulatory genomics. Regulatory encyclopedias can be created for any sequenced and annotated genome. The incorporation of such information into organism-specific regulatory repositories will provide valuable insights into the regulatory genomics.

Acknowledgments

We gratefully acknowledge support from the NSF (MCB-0418891 and MCB-0705415), DOE (DE-FG02-07ER15881) and Agricultural and Food Research Initiative Competitive Grant No. 2010-65115-20408 from the USDA National Institute of Food and Agriculture to EG and NIH Ruth L. Kirschstein National Research Service Award 5 T32 CA106196-05 from NCI to AY. Additionally, we would like to acknowledge the support of the Ohio University Stocker Endowment, Ohio University's Graduate Research and Education Board (GERB), the Ohio Supercomputer Center, the Choose Ohio First Initiative of the University System of Ohio, and the Ohio Plant Biotechnology Consortium.

References

Badawi M, Danyluk J *et al.* (2007) The *CBF* gene family in hexaploid wheat and its relationship to the phylogenetic complexity of cereal *CBF*s. *Mol Genet Genomics* 277(5): 533–554.

Davuluri R, Sun H *et al.* (2003) AGRIS: Arabidopsis Gene Regulatory Information Server, an information resource of Arabidopsis cis-regulatory elements and transcription factors. *BMC Bioinformatics* 4(1): 25.

Doherty CJ, Van Buskirk HA *et al.* (2009) Roles for Arabidopsis CAMTA transcription factors in cold-regulated gene expression and freezing tolerance. *Plant Cell* 21(3): 972–984.

Ewens WJ, Grant GR. (2001) *Statistical Methods in Bioinformatics*. Springer, New York.

Feng J, Naiman DQ *et al.* (2009) Coding DNA repeated throughout intergenic regions of the Arabidopsis thaliana genome: Evolutionary footprints of RNA silencing. *Mol BioSystems* 5(12): 1679–1687.

Higo K, Ugawa Y *et al.* (1999) Plant *cis*-acting regulatory DNA elements (PLACE) database: 1999. *Nucleic Acids Res* 27(1): 297–300.

Kaplan B, Davydov O *et al.* (2006) Rapid transcriptome changes induced by cytosolic Ca^{2+} transients reveal ABRE-related sequences as Ca^{2+}-responsive *cis* elements in Arabidopsis. *Plant Cell* 18(10): 2733–2748.

Kim SY, Chung HJ *et al.* (1997) Isolation of a novel class of bZIP transcription factors that interact with ABA-responsive and embryo-specification elements in the Dc3 promoter using a modified yeast one-hybrid system. *Plant J* 11(6): 1237–1251.

Liang X, Shen K *et al.* (2010) An integrated bioinformatics apporach to the discovery of *cis*-regulatory elements involved in plant gravitropic signal transduction. *Int J Comput Biosci* 1(1): 33–54.

Lichtenberg J, Jacox E *et al.* (2009a) Word-based characterization of promoters involved in human DNA repair pathways. *BMC Genomics* 10(Suppl 1): S18.

Lichtenberg J, Yilmaz A *et al.* (2009) The word landscape of the non-coding segments of the Arabidopsis thaliana genome. *BMC Genomics* 10(1): 463.

Morrison DR. (1968) PATRICIA — Practical Algorithm To Retrieve Information Coded in Alphanumeric. *J ACM* 15(4): 514–534.

Nakamura M, Tsunoda T *et al.* (2002) Photosynthesis nuclear genes generally lack TATA-boxes: A tobacco photosystem I gene responds to light through an initiator. *Plant J* 29(1): 1–10.

Obayashi T, Hayashi S *et al.* (2009) ATTED-II provides coexpressed gene networks for Arabidopsis. *Nucleic Acids Res* 37(suppl 1): D987–991.

Robin S, Rodolphe F *et al.* (2005) *DNA, Words and Models.* Cambridge University Press, Cambridge.

Skinner JS, von Zitzewitz J *et al.* (2005) Structural, functional, and phylogenetic characterization of a large *CBF* gene family in barley. *Plant Mol Biol* 59(4): 533–551.

Steffens NO, Galuschka C *et al.* (2004) AthaMap: An online resource for in silico transcription factor binding sites in the Arabidopsis thaliana genome. *Nucleic Acids Res* 32(suppl 1): D368–372.

Stockinger EJ, Skinner JS *et al.* (2007) Expression levels of barley *CBF* genes at the *Frost resistance-H2* locus are dependent upon alleles at *Fr-H1* and *Fr-H2*. *Plant J* 51(2): 308–321.

Thomashow, MF. (2001) So what's new in the field of plant cold acclimation? Lots! *Plant Physiol* 125(1): 89–93.

Wang J, Wong GK *et al.* (2002) RePS: A sequence assembler that masks exact repeats identified from the shotgun data. *Genome Res* 12(5): 824–831.

Welch LR, Lichtenberg J *et al.* (2009) The WORDIFIER pattern for functional and regulatory genomics. Bioinformatics Open Source Conference 2009, Stockholm, Sweden.

Wingender E, Chen X *et al.* (2000) TRANSFAC: An integrated system for gene expression regulation. *Nucleic Acids Res* 28(1): 316–319.

Chapter 8

Manycore High-Performance Computing in Bioinformatics

Jean-Stéphane Varré*,†, Bertil Schmidt‡,
Stéphane Janot*,† and Mathieu Giraud*,†

Mining the increasing amount of genomic data requires very efficient tools. The efficiency can be improved with better algorithms, but one could also take advantage of the hardware itself to reduce the application runtimes.

It has been a few years that issues with heat dissipation prevent the processors from having higher frequencies. One of the answers to maintain Moore's Law is parallel processing. Grid environments provide tools for effective implementation of coarse grain parallelization. Recently, another kind of hardware has attracted interest: the multicore processors.

Graphics Processing Units (GPUs) are a first step towards massively multicore processors. They allow everyone to have some teraflops of cheap computing power in a personal computer.

The CUDA library (released in 2007) and the new standard OpenCL (specified in 2008) make programming of such devices very convenient. OpenCL is likely to gain a wide industrial support and to become a standard of choice for parallel programming. In all cases, the best speedups are obtained when combining precise algorithmic studies with a knowledge of the computing architectures. This is especially true with the memory hierarchy: the algorithms have to find a good balance between using large (and slow) global memories and some fast (but small) local memories.

In this chapter, we will show how those manycore devices enable more efficient bioinformatics applications. We will first give some insights into architectures and parallelism. Then we will describe recent implementations specifically designed for manycore architectures, including algorithms on sequence alignment and RNA structure prediction. We will conclude with some thoughts about the dissemination of those algorithms and implementations: are they today available on the bookshelf for everyone?

---

*LIFL, UMR CNRS 8022, Université de Lille, France
†INRIA Lille Nord-Europe, France
‡School of Computer Engineering, Nanyang Technological University, Singapore

1. Introduction

Most of the algorithms presented in this book require powerful computers to run, as for example NestedMICA and R'MES tools discussed in Chapters 1 and 2. These problems are intrinsically complex to solve, and, above all, the amount of input data follows an exponential curve. With next-generation sequencers (NGS), the data produced every day are growing faster than before. The exponential grows much faster than the growth of computational power.

The challenge for computer scientists and bioinformaticians is to think about new methods and technologies which are able to deal with this incredible amount of data. A first solution is to design *better algorithms* with more elaborate data structures. However, even if a good algorithm is known, the growth of data still imposes on *better supports of execution*. To increase the computational power, one may use several computers or processors in a grid. Another possibily is to use multicore processors. Now manufacturers manage to put several processors in one: this is what is called dual-core, quad-core, or, more generally, *multicore processors.*

In this chapter, we will see how these recent advances in massively multicore processors, also called *manycore processors,* help to treat the growing flow of bioinformatics data. Manycore processors like Graphics Processing Units (GPUs) offer a huge density of computational units, for an extremely cheap price. They are more and more often directly included in personal computers, making their power available without extra cost. Thus, developments of methods using them have an impact stronger than ever. But the drawback of this technology is that the design of algorithms must take into account the constraints of these architectures. In fact, completely new distributed algorithmics are required, giving new challenges for bioinformaticians. But it is worth the effort: results published in 2008 and 2009 achieve speedups up to $100\times$ compared to serialized one-core algorithms — on commodity GPUs costing less that \$500!

In the following, we briefly describe the evolution of processors that leads to manycore processors. The evolution of power consumption and heat dissipation has led to the so-called *power wall,* that limits the rise of frequencies and commands us to use more and more cores in parallel. In Sec. 2 methods, we present manycore processors and their programming. In Sec. 3 results, we detail algorithms in different fields of bioinformatics, explaining where gains can or cannot be obtained. Finally, we discuss whether these solutions can now be used by everyone, on real applications.

1.1. *A small history of processors*

A processor, or Central Processing Unit (CPU), is roughly made of three units: the *memory unit* that stores input data, intermediate results and output data, the *instruction unit* that decodes instructions; and the *processing unit*, also called Arithmetic Logic Unit (ALU), that actually realizes the computations (Fig. 1(a)) at regular *clock cycles*. The following paragraphs explore two dimensions contributing to the "computational power" of processors: their complexity, driven by the *number of transitors*, and the *frequency* of their clock cycles.

1.1.1. *Moore's law*

The number of transistors used to build a chip has dramatically increased, from 2300 for the Intel 4004, released in 1971, to several billions today. Moore's law, formulated in (Moore, 1965) and improved in (Moore, 1975), states that this number doubles every two years. This law has

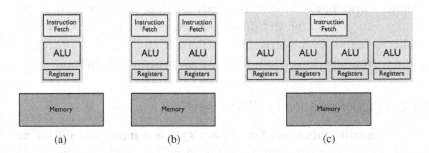

(a)   (b)   (c)

Fig. 1. (a) Von Neumann architecture: at each clock cycle, one instruction is fetched from the memory, decoded, and then executed on the ALU. Results are stored back in the registers (very fast, but small, local memory) or in the main memory (slower access). Real processors also contain one or several *caches* (not shown here) between the registers and the main memory. (b) A processor with two cores. Note that each core has its own instruction decoding, and so they can operate independently as long as they do not access the same places in the memory. (c) A SIMD processor. A single instruction decoding unit controls several processing units. Only one flow of instructions is executed, but most of the surface area of the chip is devoted to actual computation.

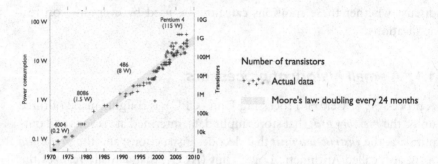

Fig. 2.   Number of transistors and power consumption of some Intel chips. Scales on the y-axis are logarithmic. Both values have grown at an exponential rate between 1970 and 2005. Data from http://www.intel.com/.

been almost exactly verified since 1960 (Fig. 2). More than a natural observation, this is a self-fulfilling prophecy that drives the semiconductor industry.

With more transistors, it is possible to increase the data width (from 4 bits in 1970 to 128 bits or more today), to build more complex operators with more complex instruction sets, and to enhance memory and caches management. This rise of the number of transistors also made possible several improvements in processor design, including instructions pipelines (dividing operators into shorter sections, processing more than one instruction at a time and thus increasing the frequency), superscalar architectures (processing several instructions at a time), out-of-order execution (permuting instructions to better use processor pipelines).

1.1.2. *Frequencies and the "power wall"*

A common misformulation of the Moore's law is that the "computational power" of the processors doubles every 18 or 24 months. Of course, all techniques cited in the previous paragraph aim to obtain more computational power. Most notably, between 1970 and 2000, the *frequencies* of the processor steadily increased, resulting in major speed gains. Between 1990 and 2000, Intel estimates that the $75\times$ gain in computational power was composed by a $13\times$ gain due to the frequencies and a $6\times$ gain due to processor design improvements (Gelsinger, 2001).

It has been a few years that issues with heat dissipation prevent the processors from having higher frequencies. Figure 2 displays some power consumptions of common CPUs. The last Intel mono-core processor, the Pentium 4, achieved a power density of about $100\,W/cm^2$. If this increase in the frequencies had continued between 2000 and 2010, the thermal density of some processors today could approach that of a rocket nozzle (Gelsinger, 2001).

1.1.3. Multicore processors

One of the answers to maintain the growth of computational power despite frequency limits is *parallel processing*. Parallelism is not new: research in distributed algorithms has been ongoing for more than 40 years with many specialized architectures for high-performance computing. Today, this field of research reaches commodity hardware through grids (several processors or computers) or multicore processors (several cores in a processor).

The idea of *multicore processors* is to have two, four, eight, or even more processors in one (Fig. 1(b)). A high majority of personal computers are provided with this kind of processors at the end of the 2000s. Basically, this can be seen as a "grid on a chip", each core having its own instruction flow. This allows to execute either the same piece of code or different instructions on the different cores, on the same data or on different data.

1.1.4. Data-parallelism and SIMD

To compute the same piece of code on different data, another idea can be implemented. Instead of several processors, we can use a single instruction unit and several processing units (Fig. 1(c)). In this case, the same piece of code is executed at the same time by several processing units. This SIMD paradigm (Single Instruction, Multiple Data) dates from the supercomputers of the 1980s. It becomes really interesting if the same task has to be executed on different data. No surface area is wasted in unnecessary control logic: the processor can be filled with efficient computing units.

1.2. Towards manycore processors

Multicore processing and SIMD processing can be combined to obtain a higher level of parallelism (Fig. 3). This is what is implemented in current *manycore* processors, including Graphical Processing Units (GPUs) and

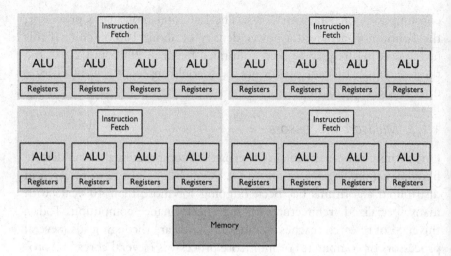

Fig. 3. A manycore processor, like current GPUs, combining several cores, each one being a high-density SIMD unit. Again, real processors also contain local memories or caches at different levels between the registers and the main memory.

Cell/BE processor. Depending on the design of the processor, emphasis is put either on the multicore or on the SIMD.

1.2.1. *GPU processors*

Today, GPUs allow everyone to have some teraflops of cheap computing power in its personal computer. High-end GPUs, for no more than $500, encompass far more arithmetic units than a CPU of the same price. For example, the NVIDIA GTX 285 has 30 cores, each one being a SIMD unit on 8 × 32 bits at about 600 MHz, whereas the ATI Radeon 4890 has 10 cores, each one being a SIMD unit on 80 × 32 bits, at about 850 MHz.

The GPUs also include memories with different sizes and capabilities. For example, the NVIDIA GTX 285 has 1 or 2 GB of *global memory*, additional global memories (constant and texture memories), and, for each core, 16 KB of *local memories*. A detailed presentation on some GPUs is available in (Fatahalian and Houston, 2008).

1.2.2. *The CPU/GPU convergence*

Recent trends blur the line between GPUs and CPUs: CPUs have more and more cores, and cores in GPUs have more and more functions. Some processors are now designed both for graphics and high-performance computing.

The Cell Broadband Engine Architecture (Gschwind *et al.*, 2006), or Cell/BE, is the chip included in the Playstation 3. It was designed by IBM and Sony and released in 2006. The Cell includes one CPU-like core, the Power Processor Element (PPE), and 8 GPU-like cores, the Synergistic Processing Elements (SPE), each one being a SIMD unit on 8×16 bits.

Larrabee (Seiler *et al.*, 2008) is an Intel graphic processor project. It can be considered as a hybrid between a CPU and a GPU. Basically, each Larrabee core is a simple Pentium core running an extended version of the $\times 86$ instruction set, but it also includes a 16×32 bits SIMD vector processing unit. The Larrabee was supposed to be released in 2010, but Intel recently announced (in December 2009) that the release is delayed, without giving any new launch date.

Intel argues that the Larrabee processor is more flexible than GPUs. However, next-generation GPUs from NVIDIA (Fermi architecture) and ATI will also have greater flexibility. In 2006, AMD (CPU manufacturer) bought ATI (GPU manufacturer), and their next project, Fusion, is also expected to mix CPU-like and GPU-like cores.

1.2.3. *General purpose computation on GPU*

Since the 1980s, graphics hardwares have been used for scientific computations (see for example (Trendall and Stewart, 2000)). The revolution concerning today's "General-Purpose Computation on Graphical Processor Units" (GPGPU) is mainly due to *the increasing number of cores*, making those chips potentially very efficient, and *the availability of programming libraries* for developers non-specialized in graphics computing (see Sec. 2 Methods). Some specialized GPUs do not have any video output: such cards are completely dedicated to general computation!

2. Methods

All manycore processors presented in the previous paragraphs have a large number of *cores*, each of them being made of a lot of processing subunits. An algorithm for such hardware thus requires a high level of *parallelism*, i.e. computations to dispatch to several cores and subunits. This section details how to program such parallelism, and the next section explains in what kind of bioinformatics applications such parallelism can be used to obtain interesting speedups.

2.1. *From GPU tweaks to OpenCL*

General-Purpose Computing on GPU, or GPGPU, was firstly done by tweaking graphics primitives. Operations on color components and pixels have been used to perform, for example, linear algebra computations. Such techniques require expertise in graphics. One of the first popular abstractions proposed for GPU programming is BrookGPU (Buck *et al.*, 2004), developed from 2003 at the Stanford University. It allows to compile and run code in the Brook Stream processing language on NVIDIA and ATI cards.

The CUDA libraries, released by NVIDIA in 2007,[a] deeply popularized GPGPU. CUDA is an extension of plain C/C++ (Fig. 4), and does not require any knowledge in graphics. Hundreds of applications in various computer science domains are now using CUDA.

The next big step was the specification of a new standard, OpenCL, by a large consortium in December 2008.[b] OpenCL is very close to CUDA, and is likely to gain a wide industrial support. OpenCL offers

```
_global__ void addv(float* a,                __kernel void addv(__global const float* a,
                    float* b,                                    __global const float* b,
                    float* c)                                    __global float *c)
{                                            {
    int i = threadIdx.x ;                        int i = get_global_gid(0);
    c[i] = a[i] + b[i] ;                         c[i] = a[i] + b[i] ;
}                                            }

main()                                       main()
{                                            {
    ...                                          ...
    addv<<< 1000, 1 >>> (a, b, c);               kernel = clCreateKernel(..., "addv", ...)
    ...                                          clEnqueueNDRangeKernel(queue, kernel, ..., 1000, ...);
}                                                ...
                                             }
```

       (a)   (b)

Fig. 4. Vector addition in (a) CUDA and (b) OpenCL. The addv function is the kernel, that is the function executed on each ALU. Each work-item executes this addition on different data, as showed by the instruction modifying the value of i. Full codes include initialization of the card, transfer of arrays a and b from the host to the GPU, and transfer of the resulting array c back to the GPU.

---

[a]http://www.nvidia.com/cuda
[b]http://www.khronos.org/opencl

a unique programming interface to deal with different manycore processors, allowing the same code to be compiled and run on different chips. But a compiler is needed for each architecture. In 2009, three different implementations, by NVIDIA, AMD/ATI and Apple, were already available, and OpenCL could become a standard for parallel programming in the following years.

In CUDA/OpenCL programs, there are some special functions, called *kernels*, that are executed on the GPU (Fig. 4). Due to the architecture of the GPU, such functions are limited: for example, as there is no stack, there are no recursive functions. Ideally, the most intensive computing tasks of an application should be mapped into a kernel. The same kernel is then executed many times by different *work-items*, working on different datasets. The *host program* calls the kernel, but also ensures that data are transferred to and back from the GPU.

2.2. *Programming SIMD work-items*

The CUDA/OpenCL kernels in Fig. 4 contain an addition instruction, $c[i] = a[i] + b[i]$. As all work-items execute this same instruction, this model is exactly the SIMD paradigm (Fig. 1(c)). This is *implicit SIMD*: the compiler generates a code that simultaneously controls all subunits.

Similar SIMD instructions are available in the common CPUs, but on a much smaller data width. For example, the original SSE instruction set, published in 1999, allows us to consider four "packed" 32-bit floats in one 128-bit machine word. The SSE assembly instruction ADDPS adds two float vectors with one single operation. This instruction can also be compiled from a C "intrinsic" instruction _mm_add_sd. Using SSE assembly or intrinsic operations is *explicit SIMD*. Even if it allows the developer to have more control, it is not as useful as CUDA or OpenCL for the programmer.

2.3. *Branches and divergence*

In the SIMD model, at one clock cycle, every arithmetic unit executes the very same instruction, allowing efficient operations on large vectors. But this is not always wanted. For example, depending on a condition in a loop, one could want to execute different instructions (Fig. 5(a)).

```
for (i=0; i<n; i++)        for (i=0; i<n; i++)
  {                          {
    if (d[i] > 53)             unsigned int mask=(d[i] > 53) ? 0xffffffff:0;
      c[i]=a[i]+b[i];          c[i] = (a[i]+b[i]) &  mask)    // then
    else                            | (d[i]        & ~mask);  // else
      c[i]=d[i];             }
  }
        (a)                                    (b)
```

Fig. 5. Simulating branching in SIMD. The branching in (a) can be simulated by the code in (b). Both branches are executed, and the mask finally chooses the good value for each item.

Such branches can be *software simulated* in SIMD extensions, by actually running both branches, and finally selecting with a mask the correct value (Fig. 5(b)). Although half of the computations are "wasted" in those branches, a good overall speedup can still be obtained. On some hardware, it is possible to temporarily disable some processing subunits into a SIMD group. This leads to *hardware simulated* branches: at a given time, only some subunits can be working in the SIMD processor (Lorie and Strong, 1984).

In CUDA/OpenCL, the developer directly writes a code similar to that shown in Fig. 5(a), and the compiler produces a code suitable for the GPU, including software or hardware simulation. When, within a same work-group, all work-items do not pass the same branch, a *divergence* occurs and some work-items are stalled (Fig. 6) (Coon and Lindholm, 2008, Collange *et al.*, 2009). Of course, a code with a lot of divergence

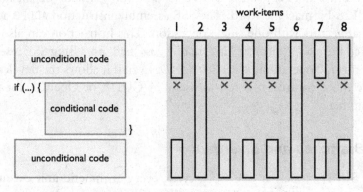

Fig. 6. Hardware simulation on branches in a SIMD unit with 8 subunits. Here only work-items 2 and 6 execute the conditional code: all other work-items are stalled, reducing the overall performance.

between work-items results in very poor performance. But, knowing this limitation, it is simpler to write such instructions in CUDA/OpenCL than to explicitly design SIMD masks.

2.4. *Work-groups*

Current GPUs are not purely SIMD, because work-items are gathered into independent *work-groups*. There is some global memory available for their communication. This *logical* organization implied by CUDA/OpenCL has different *physical* implementations in actual hardware. For example, the exact relation between the work-groups depends on the efficiency of the global memory and its caching mechanisms. On the current GPU architectures, it is advised to design truly independent work-groups, whereas the Larrabee, with global cache coherence, will allow a more traditional programming with independent threads.

3. Results

Recent parallelizations on GPUs for sequence analysis problems achieve speedups up to $100\times$ compared to a serialized one-core version. This section reviews recent results in different bioinformatics applications.

The first part of this section deals with Smith–Waterman sequence alignments. Manycore processors deliver speeds enabling us to run exact dynamic programming (DP) computations rather than incomplete heuristics. The second part explains other algorithms manipulating sequence data, including RNA algorithms and algorithms on weight matrices. Those applications use string algorithmics and DP computations similar to those used by sequence alignments, but on other objects or with different dependencies. Finally, the last part presents a few other applications where kernels deal with non-sequence data.

3.1. *Smith–Waterman sequence alignments*

In this section, we describe how the Smith–Waterman (SW) algorithm for scanning of protein sequence databases can be efficiently mapped onto some manycore architectures. Mapping onto any SIMD architecture requires choosing a fine-grained SIMD vectorization. Mapping to the Cell/BE or a GPU architecture further requires coarse-grained distribution on available cores.

3.1.1. *Smith–waterman algorithm*

SW computes the optimal local pairwise alignment (Smith and Waterman, 1981) of two given sequences S_1 and S_2 of length l_1 and l_2 using DP with the following recurrence relations.

$$\begin{cases} E(i,j) = \max\{E(i,j-1) + g_e, H(i,j-1) + g_o\} \\ F(i,j) = \max\{F(i-1,j) + g_e, H(i-1,j) + g_o\} \\ H(i,j) = \max\{0, E(i,j), F(i,j), H(i-1,j-1) + sbt(S_1[i], S_2[j])\} \end{cases},$$

where $sbt()$ is a substitution matrix such as BLOSUM62 (Henikoff and Henikoff, 1992), g_o is the gap opening penalty, g_e and is the gap extension penalty. The above recurrences are computed for $1 \le i \le l_1$ and $1 \le j \le l_2$ and are initialized as $H(i,0) = H(0,j) = E(i,0) = F(0,j) = 0$ for $0 \le i \le l_1$ and $0 \le j \le l_2$.

The score of the optimal local pairwise alignment is the maximal score in matrix H (maxScore). The actual alignment can be found by a traceback procedure. However, for SW-based protein sequence database scanning, we just need to compute (maxScore) for each query/database sequence pair. Database sequences are then ranked according to their (maxScore) value and the top hits are displayed to the user. Note that the score-only computation can be done in linear space and does not require storing the full DP matrix. The data dependency in the SW DP matrix (Fig. 7) implies that all cells in the same minor diagonal can be computed in parallel.

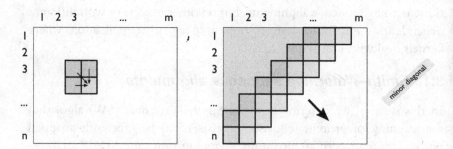

Fig. 7. Dependencies in the Smith–Waterman DP matrix. Each cell depends on its left, upper, and upper-left neighbors. All the cells in the same minor diagonal can be computed in parallel.

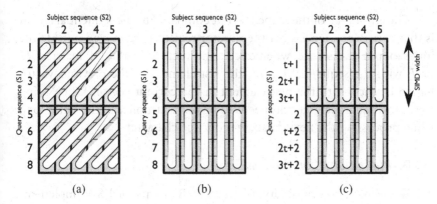

Fig. 8. SIMD vectorization approaches: (a) minor diagonal; (b) column-based with sequential layout; (c) column-based with striped layout. Each blank box depicts a parallel profile stored into a SIMD register, here with four different values.

3.1.2. *Mapping onto SIMD registers*

Between 1997 and 2007, three approaches have been proposed to vectorize SW on CPUs with SIMD instructions sets (Fig. 8): (i) *minor diagonal* approach (Wozniak, 1997), (ii) column-based approach with *sequential* memory layout (Rognes and Seeberg, 2000); (iii) column-based approach with *striped* memory layout (Farrar, 2007).

In order to calculate $H(i,j)$, the value $sbt(S_1[i], S_2[j])$ needs to be added to $H(i-1, j-1)$. The main challenge for any vectorized SW implementation is to avoid performing this table lookup for each element in a SIMD register. Therefore, all three approaches shown in Fig. 8 calculate a query profile parallel to the query sequence beforehand. In the minor diagonal approach, the query profile is then used to arrange a SIMD register with all required $sbt()$-values. The advantage of this approach is that it does not require any conditional branches in the inner loop. However, it has the disadvantages that this SIMD register has to be updated in each iteration step and that registers are not fully utilized for diagonals at the beginning and at the end. Both column-based approaches have shown higher efficiency due to the simplified dependency relationship and parallel loading of the vector scores from memory. However, a disadvantage introduced by processing in column-based order with sequential memory layout is that a conditional branch is introduced in the inner loop for computing matrix F.

The advantage of the striped layout compared to the sequential layout is that data dependencies between vector registers are moved outside the inner loop. For instance, when calculating vectors for the DP matrices H or F with the sequential layout, the last element in the previous vector has to be moved to the first element in the current vector. When using the striped query layout, this needs to be done just once in the outer loop when processing the next database sequence character.

3.1.3. *Implementation on Cell/BE*

To fully exploit the capability of the Cell/BE, a parallel SW implementation has to take advantage of the SIMD registers of each SPE. Overall, the column-based approach with striped memory layout is the most efficient of the three approaches and is consequently utilized in publicly available Cell/BE SW implementation such as SWPS3[c] (Szalkowski *et al.*, 2008) and CBESW[d] (Wirawan *et al.*, 2008). However, it should be mentioned that actual performance depends on the utilized scoring scheme (i.e. substitution matrix and gap penalties).

For coarse-grained parallelization across SPEs, the database can be partitioned into nonoverlapping workloads of similar size which are then distributed from the PPE to the different SPEs using multi-threading. Furthermore, due to the limited SPE local store (256 KB), a partitioning of the DP matrix for long query sequences is required in SWPS3 (Szalkowski *et al.*, 2008).

3.1.4. *Intra-task or inter-task parallelization on GPUs*

There are two basic approaches to map SW-based protein sequence databases scanning onto manycore GPUs: (i) inter-task and (ii) intra-task. Considering the SW alignment of a query sequence/database sequence pair as a basic task, the inter-task approach assigns each task to one work-item while the intra-task approach assigns each task to one work-group (Fig. 9). Implementation results have shown that inter-task parallelization generally achieves higher performance at the cost of higher memory consumption. Therefore, CUDASW++[e] (Liu *et al.*, 2009a) uses this method

---

[c]http://www.inf.ethz.ch/personal/sadam/swps3/
[d]http://sourceforge.net/projects/cbesw/
[e]http://cudasw.sourceforge.net/

inter-task parallelization   intra-task parallelization

Fig. 9. Inter-task vs intra-task parallelization, from (Liu *et al.*, 2009a).

for most alignments with the exception of very long database sequences, which use the intra-task approach. To achieve good load balancing within a work-group, all work-items within the same work-group can be assigned database sequences of roughly equal length.

3.1.5. *Memory optimization on GPUs*

In the CUDA programming model, global memory has to be accessed in a coalesced fashion to achieve high efficiency. Inter-task parallelization uses a coalesced global memory access pattern. A memory slot is allocated to a work-item in a work-group and is indexed top-to-bottom. Therefore, the access to the memory slot uses the same index for all work-items in a work-group. The coalesced global memory arrangement is used for both database sequences and intermediate results of the SW computation.

Fast local memory and registers can be exploited in the SW computation as follows. Instead of computing only one SW cell by each work-item for each global memory access, a whole cell block of size $n \times n$ can be computed. In this case, the computation of n cells in a column (or row) of a cell block only requires one load and one store operation to the global memory instead of n load and n store operations. Since one global access takes hundreds of clock cycles, this method leads to a significant performance improvement. However, the size of the cell block is limited by the amount of local memory and registers available; e.g. CUDASW++2.0 uses a size of 8×8 on a GTX280. Furthermore, the gap penalties and query sequence are stored in a constant global memory, while the scoring matrix is loaded to local memory.

3.1.6. *Performance comparison*

Table 1 lists some peak performance achieved by recent SW implementations, measured in GCUPS (Billion Cell Updates Per Second). In two

Table 1. Peak performance of several implementations of Smith–Waterman in GCUPS (Billion Cell Updates Per Second). All Cell/BE and GPU implementations were published in 2008 or 2009. Note that the average performance greatly depends on the input size, as well as on the score matrix and gap penalties. The 14.5 and 16.1 GCUPS of Ligowski and CUDASW++ are on dual-GPU cards.

		Peak performance	
SSEARCH (Pearson and Lipman, 1988)	CPU (2.0 GHz Xeon)	~0.1 GCUPS	non-SIMD SW implementation
(Farrar, 2007)	CPU (2.0 GHz Xeon)	3.0 GCUPS	striped layout, SSE2 SIMD instructions
CBESW (Wirawan et al., 2008)	Cell/BE	3.6 GCUPS	striped layout, long queries (800)
SWPS3 (Szalkowski et al., 2008)	Cell/BE	9.3 GCUPS	striped layout, long queries
(Farrar, 2009)	Cell/BE	16 GCUPS	striped layout, long queries (32 K)
SWCUDA (Manavski and Valle, 2008)	GeForce 8800 GTX	~1.5 GCUPS	inter-task
(Ligowski and Rudnicki, 2009)	GeForce 9800 GX2	7.5 / 14.5 GCUPS	
CUDASW++2.0 (Liu et al., 2009a)	GTX 280/295	16.8 / 28.8 GCUPS	inter-task + intra-task, short/long queries

years, this active field of research has brought a $10\times$ improvement on almost equivalent hardware. Compared to the non-SIMD plain SW implementation of Pearson and Lipman (1988), the best implementations now achieve a more than $100\times$ speedup.

However, peak performances do not measure every aspect of the implementations. For example, in (Liu et al., 2009a), the authors compare the performance of their CUDASW++ on a single-GPU GTX 280 and a dual-GPU GTX 295 versus the performance of SWPS3. SWPS3 achieves a peak performance of around 9.3 GCUPS for longer queries, but is very inefficient for short queries. Due to the inter-task parallelization, the CUDASW++2.0 single-GPU version has a relatively constant performance with an average of 16.3 GCUPS. The dual-GPU version performance is relatively low for short query sequences, but increases up to 28.8 GCUPS for longer ones.

3.2. *Algorithms on sequence data*

The algorithms discussed in this section also use kernels working on raw sequence data, but address other bioinformatics problems. The first two algorithms concern DP recurrences, whereas the last one is about weight matrices computations.

3.2.1. *RNA folding*

Determining the folding of RNAs is important for the study of noncoding RNAs. The secondary structure is a succession of *base pairings*, most often A/T, C/G and G/U. This secondary structure is a basis of the tertiary structure of the RNA. Moreover, comparative genomics has shown that, through evolution, this structure is more conserved than the nucleotide sequence.

Nussinov's algorithm finds the RNA structure with the maximum number of base pairs (Nussinov *et al.*, 1978). However, complete RNA folding algorithms are typically based on energy minimization, and include energies of stacking regions (or helices), bulge loops, internal loops, hairpin loops and multiple loops, for example as in the full Turner model (Matthews *et al.*, 1999), included in the mfold/unafold packages (Zuker, 2003).

In (Rizk and Lavenier, 2009), the authors develop an optimized GPU implementation of mfold/unafold, by computing in parallel all cells from the diagonal of the dynamic programming matrix (Fig. 10). At each cell (i, j), if bases i and j can pair, huge computations are run to search for additional RNA structures. However, in the mfold/unafold implementation, only 6 over the 16 possible combinations of bases launch these computations. That means that only 6/16 cells of the DP matrix lead to further computations. This breaks the SIMD model of the GPU, as neighbor work-items would probably diverge. To prevent this, the authors pack

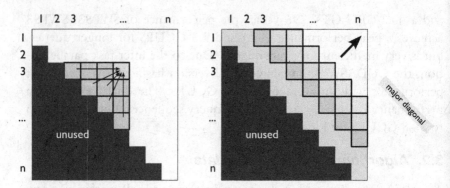

Fig. 10. Dependencies in the RNA folding DP matrix, as implemented in (Rizk and Lavenier, 2009). Each cell depends on its underlying triangle. All cells in the same major diagonal can be computed in parallel.

together these 6/16 cells, and compute all of them at the same time. Finally, they obtain a 17× speedup on a GTX 280 over a one-core version on a 2.66 GHz Xeon.

3.2.2. *Generic dynamic programming*

Algebraic Dynamic Programming (ADP) (Steffen and Giegerich, 2005) is a framework to encode different DP problems. Starting from an abstract grammar (Fig. 11) and an evaluation algebra, the ADP compiler generates the DP dependencies and recurrences, and finally translates them into C (Fig. 12). This abstraction is especially useful when the optimization problem includes lots of subcases, resulting in large DP recurrences.

```
rnafold alg f = axiom struct where
  ...
  struct     = tabulated (sadd <<< base     ~~~ struct |||
                          cadd <<< initstem ~~~ struct |||
                          nil  <<< empty ... h)

  initstem = tabulated (is <<< loc ~~~ closed ~~~ loc ... h)
  closed   = tabulated (
               stack ||| ((hairpin ||| leftB ||| rightB ||| iloop ||| multiloop) 'with' stackpairing) ... h)

  stack    = (sr  <<< base ~~~ closed ~~~ base) 'with' basepairing ... h
  hairpin  = hl  <<< base ~~~ base ~~~ (region 'with' (minsize 3))    ~~~ base ~~~ base ... h
  leftB    = bl  <<< base ~~~ base ~~~ region   ~~~ initstem  ~~~ base ~~~ base ... h
  rightB   = br  <<< base ~~~ base ~~~ initstem ~~~ region    ~~~ base ~~~ base ... h
  iloop    = il  <<< base ~~~ base ~~~ (region 'with' (maxsize 30)) ~~~ closed ~~~
                                       (region 'with' (maxsize 30)) ~~~ base ~~~ base ... h
  ...
```

Fig. 11. Excerpt from the ADP grammar for RNA folding. Starting from the axiom (**struct**), several productions detail how to generate different RNA structure components.

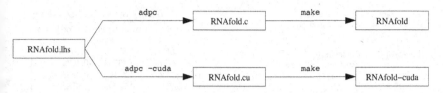

Fig. 12. ADP workflow, from (Steffen *et al.*, 2009). The same abstract grammar and algebra, in `RNAfold.lhs`, is used to produce plain C or CUDA code.

In (Steffen *et al.*, 2009), a CUDA backend of the ADP compiler is proposed.[f] In all ADP programs, all results are combined from results of shorter subsequences. Therefore, the calculation of a table element (i, j) depends only on results that lie in the "underlying triangle" under (i, j), and, as in the case of RNA folding, such elements can be computed in parallel (Fig. 10).

The authors apply this technique on several RNA tools. On pknotsRG, a RNA pseudo-knot detection application (Reeder *et al.*, 2007), they obtain speedups up to 7×. On RNA folding, they achieve a speedup up to 9×, far below the implementation of (Rizk and Lavenier, 2009). This solution is thus less efficient than a manually crafted implementation but the advantage is definitely the generic nature of the approach; even for now only a few DP problems are encoded like that.

3.2.3. *Position Weight Matrices algorithms*

Position Weight Matrices (PWMs) model approximates patterns from a set of sequences, as for instance in transcription factor binding sites, splicing sites or protein domains. Given a finite alphabet $\sum$ and a positive integer m, a PWM M is a matrix with $|\sum|$ rows and m columns (Fig. 13). The PWM associates to each word $u - u_1 u_2 \ldots u_p$ the *score* $\sum_{p=1}^{m} M(p, n_p)$. Given a score threshold α, it is possible to compute the probability to achieve a score greater than α: this is the *p*-value.

There are three usual problems on PWMs: finding positions where a pattern occurs (the *scan*), assessing a *p*-value for each occurrence, and comparing patterns. In (Giraud and Varré, 2009), a CUDA parallelization

---

[f] http://bibiserv.techfak.uni-bielefeld.de/adp/cuda.html

```
   agatcttaCGTAGTGACGTCtgccatgg
   agatctgGCGGGTGACGTGttgccatgg
   agatcttgGCGGGTGACGTTtctccatgg
              ...
 agatctcggggTATGTTGACGCCatgg
   agatCTTCGTGACGTTcttttgccatgg
   aGATCTTGACGTCcgcaggtaccatgg
```

$$M(i, x) = \log_2 \frac{\text{frequency of letter } x \text{ at position } i}{\text{background frequency of letter } x}$$

A	[-4.85826	-0.06247	-4.85826	-0.46381	···]
C	[0.37462	0.03957	-0.46793	-0.46793	···]
G	[0.03957	-0.18232	0.22107	0.82531	···]
T	[0.22314	0.22314	0.78009	-4.85826	···]

Fig. 13. A Position Weight Matrix (PWM) modeling the CREB1 transcription factor binding site (from the JASPAR database). The coefficients indicate affinities between letters and positions at the binding site: positive if the letter is over-represented in the binding site, else negative. From (Giraud and Varré, 2009).

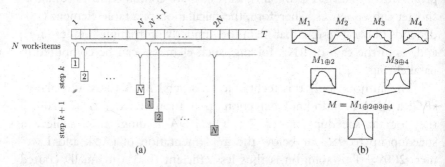

(a)   (b)

Fig. 14. (a) Parallel scan. Positions on the text τ are distributed amongst different work-items. (b) Parallel computation of PWM score distribution. The first row is computed in parallel, from brute-force word enumerations. From (Giraud and Varré, 2009).

achieves speedups of $21\times$ for the brute-force parallelization of the scan and $77\times$ for the p-value computation (on a NVIDIA GTX 280).[g]

The parallelization of the scan is straightforward, distributing positions of the text to different work-items. To reduce memory transfers, work-items in the same work-group process have interleaved positions (Fig. 14(a)).

Computing the p-value is NP-hard (Touzet and Varré, 2007). The usual algorithm proceeds by recursion (Staden, 1989), computing the score distribution Q_M, where $Q_M(s)$ is the probability to achieve exactly the score s. Then, for a given score s', the p-value is obtained with the equation: $p\text{-value}_M(s') = \sum_{s \geq s'} Q_M(s)$. The distributed algorithm of (Giraud and Varré, 2009) computes an approximation of the score distribution

---

[g]http://bioinfo.lifl.fr/cudapwm/

Q_M by splitting the matrix into submatrices. Q is computed for each submatrix and the resulting score distributions are merged (Fig. 14(b)). Here the brute-force computation of Q for each submatrix takes advantage of the GPU architecture: each work-item evaluates a large set of 4^l words (where l is the length of the submatrix), with no divergence, and no communication nor memory access during the main computation. The best speedups are obtained for larger matrices. Unfortunately, for smaller matrices, which are the ones stored into databases, the execution time remains lower using the CPU.

3.3. *Other applications*

We now briefly review other bioinformatics algorithms that have been parallelized on GPUs. Even if some problems still input sequence data, here the kernels do not directly deal with sequence data.

3.3.1. *Indexing structures*

As some string algorithms are intrinsically complex, several heuristics have been proposed on CPUs, including the popular BLAST (Altschul *et al.*, 1990) for SW sequence alignments. Heuristics often build indexes when preprocessing a part of the input. Some researchers try to put this kind of heuristics on manycore processors.

MUMmer (Kurtz *et al.*, 2004) is a heuristic that uses maximal exact matches (MEM) to seed SW sequence alignments. The MEM detection uses a large suffix tree (Gusfield, 1997). Schatz *et al.* (2007) proposed a CUDA parallelization of MUMmer, which splits the text into 8-Mb pages. Schatz and Trapnell (2009) further optimized this implementation. This latter article contains an exhaustive study of $2^7 = 128$ combinations of design choice for their CUDA implementation.[h]

Shi *et al.*, (2009) elaborated an error correction algorithm for read mapping, based on spectral alignment. Their algorithm uses a specific data structure, a Bloom filter, that is a set of hashing tables with multiple keys. The parallelization is done on the querying of this Bloom filter. They reported speedups up to $19\times$ compared to an equivalent CPU algorithm.

---

[h]http://www.mummergpu.sourceforge.net

3.3.2. *Phylogeny*

Charalambous *et al.* (2005) used the first bioinformatics application on GPU. The authors studied the RAxML phylogenetics program, and, after profiling, decided to parallelize a particular loop. The speedup ($1.2\times$) was obtained with a Brook kernel on a NVIDIA GeForce FX 5700.

Suchard and Rambaut (2009) studied statistical phylogenetics. The goal of such methods is to evaluate the likelihood of a set of sequences, given a phylogenetic tree and a model of evolution. The difficulty is in the computation of the likelihood itself. Their implementation uses very precise memory accesses, obtaining a speedup up to $60\times$ on a NVIDIA GTX 280 GPU.[i]

3.3.3. *Multiple sequence alignment*

ClustalW (Larkin *et al.*, 2007) is a multiple sequence alignment program. This progressive alignment technique aims to align the sequences following a guided tree, computed from distance matrices. Liu *et al.* (2006) proposed a parallelization of the first phase that builds pairwise alignments, with a method similar to the one presented in the first part of this section.

Liu *et al.* (2009b) elaborated a parallelization of the second phase, the Neighbor-Joining algorithm (Saitou and Nei, 1987). Here the parallelization is on the distance matrix computation, where each cell can be computed independently. The study includes a discussion on the optimal number of work-items and work-groups, as well as on a good memory management taking advantage of the symmetry of the distance matrix. A speedup up to $26\times$ is finally obtained on a NVIDIA GTX 280 card.

3.3.4. *Motif finding*

CUDA-MEME (Liu *et al.*, 2010) is a CUDA implementation of the popular MEME tool for finding regulatory regions in DNA sequences (so-called motifs). It achieves a speedup of up to $20.5\times$ over the sequential MEME code on a GTX280.

---

[i]http://beagle-lib.googlecode.com/

3.3.5. Hidden Markov Models profiles

In (Walters *et al.*, 2009), the authors propose a GPU version of HMMER (Eddy, 1998). HMMER is a popular software that aims to compute a Hidden Markov Model (HMM) from a set of aligned protein sequences, which allows searching for occurrences of similar proteins in databases. The parallelization focused on the heart of the hmmmsearch tool that uses the Viterbi algorithm. A speedup up to 38× is obtained depending on the size of the HMM.

3.3.6. Cell molecules simulation

Roberts *et al.* (2009) simulated diffusion of molecules in the cell.[j] The cell topology is modeled as a lattice, and particles follow a random walk. The speedup is not so high (2.4× on a NVIDIA GTX 280), but this study has interesting remarks on the limits of the GPU for this problem.

4. Discussion

The previous section shows that the development of manycore algorithms for bioinformatics has already started, taking into account architecture details of such processors. Here we point out what difficulties might be encountered, and discuss how to exploit the processing power of manycore processors in bioinformatics analysis. We also mention some challenges in designing parallel algorithms.

4.1. Challenges in parallel algorithmics

Most of the applications presented in this chapter exhibit *data-parallelism* on sequence data: the same instruction flow is applied to every chunk of data. This is especially the case for DP algorithms (Smith–Waterman sequence comparison, RNA folding, Algebraic Dynamic Programming), but also for other algorithms on sequence data. In all those algorithms, different work-items work on different small sections of the input data. These algorithms are the best ones to parallelize, with almost no divergence between work-items in a SIMD subunit. As mentioned in the

---

[j]http://www.ks.uiuc.edu/Research/gpu/

previous section, researchers try to further optimize those algorithms by better memory access patterns or better work-item/work-group balancing.

Another active field of research is to design elaborated data structures which are intrinsically parallel, for example, parallel structures for trees allowing large independent parallel executions. Works on suffix trees (Schatz *et al.*, 2007, Schatz and Trapnell, 2009) are a first step in this direction, but the speedups are not so high for the moment (around 5×). Techniques can be borrowed from the graphics community, where research on ray-tracing also tries to conceive good parallel algorithms for trees (Popov *et al.*, 2007).

Of course, new chips with more independent cores could allow more functionalities, but at the price of less dense computational units. Anyway, there is place for research on improved parallel algorithms, using better resources of manycore architectures.

4.2. *Challenges for bioinformatics analysis*

The end-users of algorithms and methods developed for manycore processors are supposed to be biologists or bioinformaticians who analyze data. Nevertheless, most of these programs are research prototypes rather than end-user applications. In some cases, they need better integration to be, for example, compliant with existing programs, using the same data formats for input and output.

Biomanycores. Biomanycores[k] is intended to be a collection of manycore bioinformatics tools (Varré *et al.*, 2009). The goal is both to gather manycore programs and to propose interfaces to Bio* frameworks like BioJava (Holland *et al.*, 2008), BioPerl (Stajich *et al.*, 2002), and Biopython (Cock *et al.*, 2009) (Fig. 15).

We mentioned in the previous section several parallel bioinformatics applications which are already available, especially for sequence alignment, that can provide important speedups with common GPUs. However, those programs are seldom used by, or even unknown to, biologists or bioinformaticians. Biomanycores is based on Bio* frameworks, that are widely used. These frameworks are a collection of tools that allow the user to create specific analysis pipelines for the data. With Biomanycores, the user

---

[k] http://www.biomanycores.org

```
from Bio import SeqIO
from Biomanycores import PadovaSW

bank = SeqIO.parse(open("uniprot.fa"), "fasta")
queries = SeqIO.parse(open("prot.fa"), "fasta")

for query in queries:
    handle = PadovaSW.run(query, bank)
    result = PadovaSW.SWParser().parse()
    print result
```

(a)

```
import org.biojavax.bio.seq.RichSequence;
import org.biojava.bio.dp.SimpleWeightMatrix;

import org.biomanycores.bio.pwm.*;
...
{
    RichSequenceIterator it = null;
    BufferedReader in1 = new BufferedReader(new FileReader(args[1]));
    it = RichSequence.IOTools.readFastaDNA(in1, null);
    RichSequence query = it.nextRichSequence();

    // read a weight matrix
    SimpleWeightMatrix pwm =
        PFMParser.PARSER.get(args[2], alph, "ACGT");

    // scan the sequence
    LillePWMScan scanner = new LillePWMScan(launcher);
    List<PWMHit> al = scanner.scan(query,pwm,2500.0);
    ...
}
```

(b)

Fig. 15. Biomanycores interfaces. (a) Biopython code for a SW comparison. (b) BioJava code for a computation with a weight matrix. The interfaces try to use as far as possible the standard objects of existing APIs (such as Biopython's `SeqIO`).

can keep the pipeline and increase the capacity of analysis by replacing the sequential tool by a parallel one. For example, suppose that a biologist has a large number of sequences to analyze, and wants to format the results. By simply replacing the call to the alignment function by a call to the parallel one, he/she can have a speedup of around $10\times$. Concretely, Biomanycores is a repository of open-source parallel bioinformatics code (in Opencl or Cuda) and interfaces to use these programs from Bio* frameworks. It tries to bridge the gap between research in high-performance computing and platforms used by bioinformaticians and biologists, by giving access to high-performance prototypes through Bio* frameworks. The project started with three different applications: Smith–Waterman (Manavski and Valle, 2008), pKnotsRG and weight matrices scan (see Sec. 3 Results). Contributors are welcome to upload their own applications. Of course, the speedup using those APIs is lower than the one obtained by directly using the applications or a manually crafted pipeline, but the advantage is the ease of use inside standard frameworks.

5. Conclusion

Initiated by multicore and graphics processors, the trend of having more cores in a processor will surely continue in the coming years. As explained in the introduction, in the future, the increase of computational power will come from a higher number of processors (either in CPUs or in GPUs) rather than higher frequencies. On the other hand, the amount of biological data produced is growing more and more rapidly, and this increases the need for high-performance bioinformatics applications. Parallel algorithms could then become the standard, since parallelism is necessary to fully exploit the power of such processors. New tools like CUDA or OpenCL, as detailed in Sec. 2, makes the use of GPUs for general programming easier, allowing the user to benefit from the computational power of GPUs.

In the third section, we have studied some parallel bioinformatics applications. Parallelism is particularly suitable for bioinformatics algorithms based on dynamic programming, like sequence comparisons. Those applications can achieve good speedups, even on common (and cheap) GPUs. But the use of parallelism is not always straightforward, and more work is needed, for example, to design parallel data structures or to find new parallel methods. Bringing parallelism to biologists and bioinformaticians is another challenge; the Biomanycores project is a first step toward this goal.

We believe that the development of manycore processors (on commodity hardware) starts a new era in parallel processing, and creates opportunities for high-performance computing in bioinformatics.

References

Altschul SF, Gish W, Miller W, Myers EW, Lipman DJ. (1990) Basic local alignment search tool. *J Mol Biol* **215**(3): 403–410.

Buck I, Foley T, Horn D *et al.* (2004) Brook for GPUs: Stream computing on graphics hardware. *ACM Transactions on Graphics* **23**: 777–786.

Charalambous M, Trancoso P, Stamatakis A. (2005) Initial experiences porting a bioinformatics application to a graphics processor. *Advances in Informatics*, 415–425.

Cock PJ, Antao T, Chang JT *et al.* (2009) Biopython: Freely available Python tools for computational molecular biology and bioinformatics. *Bioinformatics* **25**(11): 1422–1423.

Collange S, Daumas M, Defour D, Parello D. (2009) Comparaison d'algorithmes de branchements pour le simulateur de processeur graphique Barra. In: *RenPar' 19/SympA' 2009/CFSE' 7.*

Coon BW, Lindholm JE. (2008) *System and Method for Managing Divergent Threads in a SIMD Architecture.* US patent 7353369.

Eddy SR. (1998) Profile Hidden Markov Models. *Bioinformatics* **14**: 755–763.

Farrar M. (2007) Striped Smith–Waterman speeds database searches six times over other SIMD implementations. *Bioinformatics* **23**(2): 156–161.

Farrar MS. (2009) *Optimizing Smith–Waterman for the Cell Broadband Engine.* http://farrar.michael.googlepages.com/smith-watermanfortheibmcellbe.

Fatahalian K, Houston M. (2008) A closer look at GPUs. *Commun ACM* **51**(10): 50–57.

Gelsinger P. (2001) Microprocessors for the new millennium: Challenges, opportunities, and new frontiers. *IEEE International Solid State Circuits Conference (ISSCC 2001)*, pp. 22–25.

Giraud M, Varré JS. (2009) Parallel position weight matrices algorithms. In: *International Symposium on Parallel and Distributed Computing (ISPDC 2009)*.

Gschwind M, Hofstee HP, Flachs B, Hopkins M, Watanabe Y, Yamazaki T. (2006) Synergistic processing in cell's multicore architecture. *IEEE Micro* **26**(2): 10–24.

Gusfield D. (1997) *Algorithms on Strings, Trees, and Sequences.* Cambridge University Press.

Henikoff J, Henikoff S. (1992) Amino acid substitution matrices form protein blocks. *Proc Natl Acad Sci USA* **89**: 10915–10919.

Holland RCG, Down TA, Pocock M *et al.* (2008) BioJava: An open-source framework for bioinformatics. *Bioinformatics* **24**(18): 2096–2097.

Kurtz S, Phillippy A, Delcher AL *et al.* (2004) Versatile and open software for comparing large genomes. *Genome Biol.* **5**(R12).

Larkin M, Blackshields G, Brown N *et al.* (2007) Clustal W and Clustal X version 2.0. *Bioinformatics* **23**(21): 2947–2948.

Ligowski L, Rudnicki W. (2009) An efficient implementation of Smith–Waterman algorithm on GPU using CUDA, for massively parallel scanning of sequence databases. In: *IEEE International Workshop on High Performance Computational Biology (HiCOMB 2009)*.

Liu W, Schmidt B, Voss G, Müller-Wittig W. (2006) GPU-ClustalW: Using graphics hardware to accelerate multiple sequence alignment. *IEEE International Conference on High Performance Computing (HiPC 2006)*. Lecture Notes in Computer Science (LNCS), vol. 4297, pp. 363–374.

Liu Y, Schmidt B, Liu W, Maskell D. (2010) CUDA-MEME: Accelerating motif discovery in biological sequences using CUDA-enabled graphics processing units. *Pattern Recognition Letters*. In press.

Liu Y, Maskell D, Schmidt B. (2009a) CUDASW++: Optimizing Smith–Waterman sequence database searches for CUDA-enabled graphics processing units. *BMC Research Notes* **2**(1): 73.

Liu Y, Schmidt B, Maskell D. (2009b) Parallel reconstruction of neighbor-joining trees for large multiple sequence alignments using CUDA. In: *IEEE International Workshop on High Performance Computational Biology (HiCOMB 2009)*.

Lorie RA, Strong HR Jr. (1984) *Method for Conditional Branch Execution in SIMD Vector Processors.* US patent 4435758.

Manavski SA, Valle G. (2008) CUDA compatible GPU cards as efficient hardware accelerators for Smith–Waterman sequence alignment. *BMC Bioinformatics* **9** Suppl 2: S10.

Matthews DH, Sabrina J, Zuker M, Turner DH. (1999) Expanded sequence dependence of thermodynamic parameters improves prediction of RNA secondary structure. *J Mol Biol* **288**: 911–940.

Moore GE. (1965) Cramming more components onto integrated circuits. *Electronics* **38**(8).

Moore GE. (1975) Progress in digital integrated electronics. *Technical Digest 1975. International Electron Devices Meeting, IEEE* pp. 11–13.

Nussinov R, Pieczenik G, Griggs J, Kleitman D. (1978). Algorithms for loop matchings. *SIAM J Appl Math* **35**: 68–82.

Pearson W, Lipman D. (1988) Improved tools for biological sequence comparison. *Proc Natl Acad Sci* **85**: 3244–3248.

Popov S, Günther J, Seidel HP, Slusallek P. (2007) Stackless KD-tree traversal for high performance GPU ray tracing. *Computer Graphics Forum* **26**(3): 415–424.

Reeder J, Steffen P, Giegerich R. (2007) pknotsRG: RNA pseudoknot folding including near-optimal structures and sliding windows. *Nucleic Acids Res* **35**(S2): W320–324.

Rizk G, Lavenier D. (2009) GPU accelerated RNA folding algorithm. In: *Using Emerging Parallel Architectures for Computational Science / International Conference on Computational Science (ICCS 2009)*.

Roberts E, Stone J, Sepulveda L, Hwu WM, Luthey-Schulten Z. (2009) Long time-scale simulations of *in vivo* diffusion using GPU hardware. In: *IEEE International Workshop on High Performance Computational Biology (HiCOMB 2009)*.

Rognes T, Seeberg E. (2000) Six-fold speed-up of Smith–Waterman sequence database searches using parallel processing on common microprocessors. *Bioinformatics* **16**(8): 699–706.

Saitou N, Nei M. (1987) The neighbor-joining method: A new method for reconstructing phylogenetic trees. *Mol Biol Evol* **4**(4): 406–425.

Schatz MC, Trapnell C, Delcher AL, Varshney A. (2007) High-throughput sequence alignment using Graphics Processing Units. *BMC Bioinformatics* **8**: 474.

Schatz MC, Trapnell C. (2009) Optimizing data intensive GPGPU computations for DNA sequence alignment. *Parallel Computing* **35**: 429–440.

Seiler L, Carmean D, Sprangle E, *et al.* (2008) Larrabee: A many-core x86 architecture for visual computing. *ACM Trans Graphics* **27**(3).

Shi H, Schmidt B, Liu W, Mueller-Wittig W. (2009) Accelerating error correction in high-throughput short-read DNA sequencing data with CUDA. In: *IEEE International Workshop on High Performance Computational Biology (HiCOMB 2009)*.

Smith TF, Waterman MS. (1981) Identification of common molecular subsequences. *J Mol Biol* **147**: 195–197.

Staden R. (1989) Methods for calculating the probabilities of finding patterns in sequences. *Computer Appl Biosci* **5**(2): 89–96.

Stajich JE, Block D, Boulez K *et al.* (2002) The Bioperl toolkit: Perl modules for the life sciences. *Genome Res* **12**(10): 1611–1618.

Steffen P, Giegerich R. (2005) Versatile and declarative dynamic programming using pair algebras. *BMC Bioinformatics* **6**(1).

Steffen P, Giegerich R, Giraud M. (2009) GPU Parallelization of algebraic dynamic programming. In: *Parallel Processing and Applied Mathematics/Parallel Biocomputing Conference (PPAM/PBC 09)*.

Suchard MA, Rambaut A. (2009) Many-core algorithms for statistical phylogenetics. *Bioinformatics* **25**(11): 1370–1376.

Szalkowski A, Ledergerber C, Krahenbuhl P, Dessimoz C. (2008) SWPS3 — fast multi-threaded vectorized Smith–Waterman for IBM Cell/B.E. and x86/SSE2. *BMC Research Notes* **1**(1): 107.

Touzet H, Varré JS. (2007) Efficient and accurate P-value computation for Position Weight Matrices. *Algorithms Mol Biol* **2**(1).

Trendall C, Stewart AJ. (2000) General calculations using graphics hardware with applications to interactive caustics. *Proceedings of the Eurographics Workshop on Rendering Techniques 2000*, pp. 287–298, Springer-Verlag London.

Varré JS, Janot S, Giraud M. (2009) Biomanycores, a repository of interoperable open-source code for many-cores bioinformatics. In: *Bioinformatics Open Source Conference (BOSC)*.

Walters JP, Balu V, Kompalli S, Chaudhary V. (2009) Evaluating the use of GPUs in liver image segmentation and HMMER database searches. In: *IEEE Int Symp Parallel and Distributed Processing*, Springer, pp. 1–12.

Wirawan A, Kwoh CK, Nim TH, Schmidt B. (2008) CBESW: Sequence alignment on the Playstation 3. *BMC Bioinformatics* **9**(377).

Wozniak A. (1997) Using video-oriented instructions to speed up sequence comparison. *Computer Appl Biosci* **13**: 145–150.

Zuker M. (2003) Mfold web server for nucleic acid folding and hybridization prediction. *Nucleic Acids Res* **31**(13): 3406–3415.

Chapter 9

Natural Selection and the Genome

Austin L. Hughes*

1. Introduction

In 2009, the 200th anniversary of Charles Darwin's birth and the 150th anniversary of his publication of *On the Origin of Species* were marked by dozens of symposia, books, and articles, both scholarly and popular. Amid this acclamation, we heard frequent references to "Darwinism" — a kind of terminology more appropriate for a political (e.g. "Marxism"), philosophical (e.g. "Platonism"), or religious (e.g. "Calvinism") doctrine than for a scientific theory. Critical assessments of Darwin's contribution to biology were few, and the opportunity to assess exactly how far evolutionary biology has progressed since Darwin's seminal publication was largely missed.

It is no detraction from Darwin's contribution to biology to point out that his understanding of evolution was rather limited by modern standards, and that the field has progressed greatly since his time. Indeed, it is a tribute to Darwin's commitment to a rigorous scientific approach to questions that, before his time, were most often addressed in a highly speculative, philosophical way. Because of the solid foundation given to the field of evolutionary biology by Darwin and several of his contemporaries, they set in motion a process leading eventually to discoveries that rendered their own views obsolete. This is how science is supposed to work — in contrast to philosophical dogmatism.

*Department of Biological Sciences, University of South Carolina, Columbia, SC 29208, USA

2. The Molecular Revolution

It might be argued that most of the progress that has occurred in evolutionary biology since its inception has occurred in just the past 30 years or so; that is, in the short time since the availability of nucleotide sequence data made possible the study of evolution at the most fundamental level. Though phenotypic changes are the consequence of evolutionary change, when biologists were confined to the study of phenotypes, they were not really able to study evolution itself. Evolution in its essence is change at the nucleotide sequence level, and without access to the genotype, evolutionary biologists were always one step away from where the real action is. As long as the molecular basis of phenotypes remained unknown, it was essentially impossible to understand their evolution.

When Darwin published *On the Origin of Species* in 1859, not only was evolution not a widely accepted idea even among scientists, but also the mechanism of natural selection proposed by Darwin was novel. Thus, there was a tendency to confound the hypothesis of evolution with the proposed mechanism. In a number of cases in his writings, Darwin referred to "my theory" without making it clear whether he meant evolution itself or natural selection. Unfortunately, the tendency to confound evolution and natural selection persists to this day, when it is no longer necessary or fruitful.

Today no scientist would question the statement that life on earth is the result of an evolutionary process. Moreover, we have abundant evidence that natural selection has occurred in the past and is occurring in contemporary populations. But today we also have a great deal of evidence suggesting that the kind of natural selection envisioned by Darwin is by no means the only factor at work in the evolutionary process. Darwin focused mainly on what today is known as positive selection — or "positive Darwinian selection" — that is, selection favoring adaptive mutations. But the molecular evidence suggests that positive selection is very rare and sometimes rather trivial in its effects. Rather, the molecular evidence overwhelmingly identifies purifying selection — that is, selection acting against deleterious mutations — as the predominant form of natural selection that occurs in nature.

The over-emphasis on positive selection is a legacy of the period of Neo-Darwinism. Neo-Darwinism arose in the 1920s and 1930s, when modern population genetics was developed by biologists such as Fisher,

Haldane, and Wright in the attempt to synthesize Darwin's theory of natural selection with Mendelian genetics. The Neo-Darwinists, especially Fisher and his school, postulated that new adaptive traits arise mainly (if not exclusively) according to a scenario by which a new advantageous mutation occurs on a single chromosome in a population and eventually is driven to fixation (i.e. a frequency of 100%) by positive selection. In the molecular era, this scenario imposes an outmoded straight-jacket on evolutionary thinking that has harmful consequences for the biological sciences, including their application to human health.

3. The Neutral Theory

When the first amino acid sequence data became available in the 1960s, along with the first data on the level of polymorphism at protein-coding loci, both Kimura (1968) and Jukes and King (1969) proposed that much of the polymorphism observed in natural populations is selectively neutral and that genetic drift is the primary mechanism leading to fixation of new variants at the molecular sequence level. Kimura's so-called neutral theory of molecular evolution, which was grounded in his intensive previous study of the stochastic element in population genetics, was controversial in the 1970s, giving rise to the selectionist-neutralist debate. For a summary of Kimura's theory and the main issues in the debate, see Kimura (1983).

In the 1980s, the advent of rapid DNA sequencing somewhat altered the terms of the debate. Sequencing revealed the existence of noncoding introns breaking up the protein-coding sequence of many eukaryotic genes; moreover, it was discovered that the vast majority of the genome of many eukaryotes is noncoding, often consisting of repeats and other apparently functionless DNA. As a consequence, even many staunch selectionists were willing to concede that mutations in noncoding regions may often be selectively neutral and that genetic drift thus must determine their fate in populations. At the same time, most neutralists were willing to concede that positive selection plays an important role in evolution, even though it is responsible for only a minority of fixation events.

Moreover, the advent of abundant sequence data provided strong support for Kimura's prediction that purifying selection, not positive selection, predominates at the molecular level. An important source of insights into the role of natural selection was provided by the redundancy of the genetic code, which makes it possible to estimate separately the number of synonymous nucleotide substitutions per synonymous site (d_S) and the number of nonsynonymous nucleotide substitutions per nonsynonymous site (d_N). Nonsynonymous mutations, because they change the amino acid sequence, are generally harmful to protein structure. Thus, many nonsynonymous mutations will be quickly eliminated by purifying selection. Other nonsynonymous mutations, which are selectively neutral or very slightly deleterious, may persist in populations and eventually become fixed by genetic drift. Synonymous mutations, because they do not change the amino acid sequence, are generally much less likely to be subject to purifying selection. As a consequence, it is predicted by the neutral theory that d_S will exceed d_N in most proteins; and support for this prediction has provided strong support for the neutral theory (Li et al., 1985).

The central insight behind such comparisons can readily be extended outside coding regions. For example, in bacteria, the rate of substitution in spacers between genes has been shown to be significantly reduced in comparison to that of synonymous substitution in the adjacent gene (Hughes and Friedman, 2004). Moreover, the search for potentially functionally important regions of the genome outside coding regions is generally conducted by looking for regions of conservation between species, using sequence alignment programs (MultiPipmaker (Schwartz et al., 2003), rVISTA (Loots and Ovcharenko, 2004)). That conservation is taken as evidence of important function, based on the insight of the neutralists that natural selection is primarily purifying.

4. Positive Selection: The MHC Case

Comparison of d_S and d_N has also been used to study exceptional cases of positive selection at the molecular sequence level, e.g. the highly polymorphic genes of the major histocompatibility complex (MHC) genes of vertebrates (Hughes and Nei, 1988; 1989). The polymorphism of the MHC genes was discovered — because of the role of the class I MHC genes in transplant rejection — before the function of these genes was known. As a result, biologists proposed numerous hypotheses to account for both MHC polymorphism and function, some of which were quite fanciful. For example, since the only other genetic system known to display a level of polymorphism equal to that of the MHC was the self-incompatibility system of plants, it was proposed that the MHC is a kind of self-incompatibility system for vertebrates (Thomas, 1974). Amazingly, an extensive literature continues to be produced regarding the supposed role of the MHC in mate-choice in a variety of species (including humans), despite the fact that no adequately controlled study has ever demonstrated such a phenomenon (Hughes and Yeager, 1998).

Zinkernagel and Doherty (1974) unraveled the mystery of class I MHC function, showing that these molecules function to present peptides to cytotoxic T-cells (CTL), which function to kill cells infected by intracellular pathogens such as viruses. The same authors soon provided evidence that different class I MHC allelic products bind different peptides (Doherty and Zinkernagel, 1975). The latter discovery suggested the hypothesis that MHC polymorphism is maintained by overdominant selection (heterozygote advantage). Since a heterozygote at a given MHC locus will be able to bind and present to CTL a broader array of pathogen-derived peptides, heterozygosity should be predicted to provide an advantage in terms of disease resistance (Doherty and Zinkernagel, 1975).

Working shortly after the first crystal structure of a class I MHC molecule had been reported (Bjorkman *et al.*, 1987), Hughes and Nei (1988) predicted that, if Doherty and Zinkernagel's (1975) hypothesis is correct, amino acid changes should be selectively favored in the peptide-binding region (PBR) of the class I MHC molecule. Thus, in the codons encoding the PBR, we might expect to see the highly unusual pattern $d_N > d_S$, whereas in the rest of the gene we should see $d_S > d_N$, as in most genes. This prediction was supported, in both class I and class II MHC genes, consistent with the predictions of Doherty and Zinkernagel's (1975) hypothesis.

The MHC provides an example of balancing selection, i.e. selection that acts to maintain a polymorphism (of which overdominant selection is the best-understood mechanism). Balancing selection is one of two kinds of positive selection; the other is directional selection, which is selection leading to fixation of an advantageous variant. Note that Darwin himself focused on directional selection, being unaware of balancing selection. The neutral theory predicts that both forms of positive selection are rare, and data from sequence analyses overwhelmingly support this prediction. However, the literature has been increasingly clogged over the past decade or so with fallacious claims of positive selection at the molecular level, based on the use of poorly conceived statistical methods (Hughes, 2007). In order to appreciate the state of evolutionary biology today, it is important that we understand why the reasoning behind these statistical methods is faulty, and the consequent serious damage to evolutionary biology as a science.

5. Codon-Based Methods

In the MHC case, we were testing a specific prediction of an *a priori* biological hypothesis; namely, that if selection favored diversity in the PBR because it confers enhanced immune surveillance, multiple amino acid substitutions should be selectively favored in the PBR (Hughes and Nei, 1988; 1989). It does not follow that positive selection must be occurring in every case where $d_N > d_S$ in an individual codon or set of codons. Rather, it can easily be shown that such codons will occur in virtually every dataset of aligned coding sequences, even under strong purifying selection, due to the random nature of the mutational process (Hughes and Friedman, 2005; 2008). Yet, many computational biologists have made the unwarranted assumption that the existence of even a single codon with $d_N > d_S$ constitutes a "signature" of positive selection.

This false assumption provides the basis of the so-called "codon-based" tests for positive selection (Hughes, 2007). Because these methods identify as "positively selected" *any* codon with one or more nonsynonymous changes and no synonymous changes (Hughes and Friedman, 2008), they are highly sensitive to sequencing and alignment errors (Schneider *et al.*, 2009). Although thousands of papers have been published based on these methods, none of the alleged cases of positive selection which were supposedly uncovered can be taken seriously. A recent study by Yokoyama and colleagues (2008) provides dramatic evidence of how misleading these methods are. On the basis of both phylogenetic analysis and laboratory experiments, these authors were able to identify the amino acid changes underlying adaptive differences among vertebrate rhodopsins (Yokoyama *et al.*, 2008). On the other hand, the "codon-based" methods completely failed to identify the amino acids actually involved in adaptive evolution, focusing instead on irrelevant changes.

6. The McDonald–Kreitman Test

An equally widespread and equally misguided method is the McDonald–Kreitman test (McDonald and Kreitman, 1991), of which there are several versions and modification in use. The basis idea of the McDonald–Kreitman test is to compare the ratio of nonsynonymous polymorphisms (Pn) to synonymous polymorphisms (Ps) in a species with the ratio of nonsynonymous (Dn) to synonymous (Ds) substitutions between that species and an outgroup species. The expectation under strict neutrality is that $Pn:Ps$ will equal $Dn:Ds$. Typically cases where $Dn:Ds$ significantly exceeds $Pn:Ps$ are taken by users of the method to indicate positive selection favoring fixation of amino acid differences between species.

The problem with this method is that a significant result may have more than one interpretation besides positive selection. Several alternative explanations involve slightly deleterious mutations. When effective population size is small, purifying selection is ineffective in removing slightly deleterious mutations, thus allowing them to increase in frequency or even become fixed as a result of genetic drift (Ohta, 1973). Slightly deleterious mutations in coding regions are much more likely to be nonsynonymous than synonymous. If a bottleneck occurs during speciation, the result may be an increase in $Dn:Ds$ and thus a significant result of the test. This phenomenon seems to have occurred in the case of the alcohol dehydrogenase (*Adh*) gene of Hawaiian *Drosophila* (Ohta, 1993).

A deeper problem with the McDonald–Kreitman test affects even species that have not undergone bottlenecks. All populations of organisms contain numerous slightly deleterious variants, and those that occur in coding regions are largely nonsynonymous. However, the intensity of purifying selection against such variants will vary across loci. At loci where this selection is very intense, Pn will be low relative to Ps. As a consequence, at these loci, $Dn:Ds$ will appear high relative to $Pn:Ps$. Those who use the McDonald–Kreitman test uncritically will conclude that positive selection is occurring between species when in fact particularly strong purifying selection is occurring within species (Hughes *et al.*, 2007). Thus, this test mistakenly identifies the very proteins that are most constrained as subject to positive selection!

The McDonald–Kreitman test has sometimes been extended to areas of the genome outside protein-coding genes (e.g. Andolfatto, 2005). In this case, polymorphism and divergence in noncoding regions replace those at nonsynonymous sites, with synonymous sites still being used as a neutral standard. Here, the inherently poor design of the test also renders the supposed "signal" of positive selection essentially meaningless, as even synonymous sites within a coding sequence can be subject to purifying selection due to various functional constraints (e.g. taking part in *cis*-regulatory regions). Again effective purifying selection within species is mistaken for positive selection between species.

7. Single Nucleotide Polymorphisms

With the accumulation of data on single nucleotide polymorphisms (SNPs) in humans and other organisms, one striking observation has been found in virtually every species: there are numerous nonsynonymous SNPs in coding regions, but nonsynonymous SNPs tend to have lower gene diversities (heterozygosities) than synonymous SNPs in the same genes (Hughes, 2005; Hughes *et al.*, 2003). Since most human SNPs are bi-allelic (reflecting the overall low genetic diversity in humans), one way that diversity at human SNPs is often measured is by the "minor allele frequency" (MAF), the frequency of the less common allele. MAF is negatively correlated with gene diversity. Thus, MAF at nonsynonymous SNPs tend to be lower than MAF at synonymous SNPs in the same genes.

The most obvious interpretation of this finding is that most of the nonsynonymous SNPs in the human genome represent slightly deleterious variants that are subject to ongoing purifying selection (Hughes *et al.*, 2003; Sunyaev *et al.*, 2001; Zhao *et al.*, 2003). The effects of purifying selection explain the reduction in MAF at nonsynonymous SNP sites. Note that the MAF values at these nonsynonymous SNP loci are generally in the 1–10% range. These values are thus much higher than MAFs at classic Mendelian disease genes (such as the genes causing cystic fibrosis or Huntington's chorea), which are kept at very low frequency (typically 10^{-3} or less) in a mutation-selection balance maintained by strong purifying selection. The existence of numerous slightly deleterious variants in the human population is consistent with the bottlenecked population history of humans, since in a small population purifying selection is not effective in removing slightly deleterious variants (Hughes *et al.*, 2003).

Given the abundance of SNP data from humans, there has been substantial interest in finding supposed "signatures" of positive selection in SNP data. As with other so-called tests for positive selection, the methods that have been proposed to identify "signatures" of positive selection in

SNP data are all problematic; thus, in spite of the fact that these analyses have been reported in numerous high-profile publications, few of them can be taken seriously. An example of such a defective method is a test based on MAF (Walsh *et al.*, 2006). This test is based on the reasoning that a "selective sweep" will cause a reduction in MAF at the selected site and linked sites, while balancing selection will cause high MAF. A selective sweep is hypothesized to occur as a result of recent fixation or near-fixation of some selectively advantageous variant, leading to a reduction in diversity in sites closely linked to the site under selection.

However, there are other population processes besides a selective sweep that can cause a reduction in MAF at a number of linked sites. Likewise, there are factors other than balancing selection that can cause high MAF at a number of linked sites. The most obvious factor in both cases is genetic drift. Both of these patterns will occur in a number of genomic regions by chance as a result of genetic drift. Population bottle-necks — as occurred when the ancestors of the European and Asian human populations left Africa — will greatly enhance the likelihood of dramatic changes in haplotype frequency as a result of genetic drift.

Moreover, because the human genome is subject to ongoing purifying selection against numerous slightly deleterious variants, it is not surprising that purifying selection will sometimes lead to reduced MAF at a number of linked sites. Thus, the test based on MAF has the same problem as other supposed tests of positive selection mentioned previously, i.e. that it cannot distinguish positive selection from purifying selection. It has the added disadvantage that it cannot even distinguish positive selection from genetic drift.

Another type of test that has similar problems is one based on "derived allele frequency" (DAF), again proposed by Walsh *et al.* (2006). Using nonhuman primate sequences as an outgroup, it is possible to determine in many cases whether one or both of the alleles at a human bi-allelic SNP site is ancestral and which is derived. Walsh *et al.* (2006) argue that very high or very low DAF might indicate positive selection. However, since evolution is generally conservative, it might be predicted that derived alleles are more likely to be deleterious than ancestral alleles are. Indeed, there is evidence that this is so in the human genome, since the derived allele tends to be the rarer allele at human nonsynonymous SNPs (Hughes *et al.*, 2003).

8. Conclusions: The Importance of Purifying Selection

Misled by the outmoded concepts of Neo-Darwinism, many researchers have sought evidence of positive selection with little reflection on why knowledge of positive selection would be useful or important, or if the supposed evidence were even reliable. Sometimes such searches are justified by the claim that positive selection on a gene should reveal a role in disease. This frequently repeated idea reveals a failure to grasp the most fundamental concepts of biology. A positively selected variant is one that confers an advantage, and therefore ordinarily should not cause disease. It is true that, in the case of the sickle-cell gene, overdominant selection has actually greatly increased in frequency a disease gene. But this is surely a very unusual case, resulting from the very strong selection pressure imposed by virulent malaria.

A result of the obsession with positive selection is that purifying selection has been comparatively neglected. Yet purifying selection is by far the most important form of natural selection, both with regard to prevalence and with regard to what it can tell us about biological function. Moreover, the fact that purifying selection is not always effective (depending on population size) causes slightly deleterious mutations to be abundant in the human population (and in many other populations). These slightly deleterious variants include not only nonsynonymous SNPs in coding regions, but also certain SNPs outside of coding regions; for example, in regulatory sequences and perhaps in microRNAs. It is these abundant slightly deleterious mutations that should be our focus if we are trying to determine the causes of complex human diseases such as heart disease and cancer (Hughes *et al.*, 2003).

References

Andolfatto P. (2005) Adaptive evolution of non-coding DNA in *Drosophila*. *Nature* **437**: 1149–1153.

Bjorkman PJ, Saper MA, Samraoui B, Bennett WS, Strominger JL, Wiley DC. (1987) The foreign antigen binding site and T cell recognition regions of class I histocompatibility antigens. *Nature* **329**: 512–518.

Hughes AL. (2005) Evidence for abundant slightly deleterious polymorphisms in bacterial populations. *Genetics* **169**: 533–538.

Hughes AL. (2007) Looking for Darwin in all the wrong places: The misguided quest for positive selection at the nucleotide sequence level. *Heredity* **99**: 364–373.

Hughes AL, Friedman R. (2004) Patterns of sequence divergence in 5' intergenic spacers and linked coding regions in 10 species of pathogenic Bacteria reveal distinct recombinational histories. *Genetics* **168**: 1795–1803.

Hughes AL, Friedman R. (2005) Variation in the pattern of synonymous and nonsynonymous difference between two fungal genomes. *Mol Biol Evol* **22**: 1320–1324.

Hughes AL, Friedman R. (2008) Codon-based tests of positive selection, branch lengths, and the evolution of mammalian immune system genes. *Immunogenetics* **60**: 495–506.

Hughes AL, Friedman R, Rivailler P, French JO. (2008) Synonymous and nonsynonymous polymorphisms and divergences in bacterial genomes. *Mol Biol Evol* **25**: 2199–2209.

Hughes AL, Nei M. (1988) Pattern of nucleotide substitution at MHC class I loci reveals overdominant selection. *Nature* **355**: 402–403.

Hughes AL, Nei M. (1989) Nucleotide substitution at at major histocompatibility complex class II loci: Evidence for overdominant selection. *Proc Natl Acad Sci USA* **86**: 958–962.

Hughes AL, Packer B, Welch R, Bergen AA, Chanock SJ, Yeager M. (2003) Widespread purifying selection at polymorphic sites in human protein-coding loci. *Proc Natl Acad Sci USA* **100**: 15754–15757.

Hughes AL, Yeager M. (1998) Natural selection at major histocompatibility complex loci of vertebrates. *Annu Rev Genet* **32**: 415–435.

Jukes JL, King TH. (1969) Non-Darwinian evolution. *Science* **164**: 788–798.

Kimura M. (1968) Evolutionary rate at the molecular level. *Nature* **217**: 624–626.

Kimura M. (1983) *The Neutral Theory of Molecular Evolution*. Cambridge University Press, Cambridge.

Li W-H, Wu C-I, Luo C-C. (1985) A new method for estimating synonymous and nonsynonymous rates of nucleotide substitution considering the relative likelihood of nucleotide and codon changes. *Mol Biol Evol* **2**: 150–174.

Loots GG, Ovcharenko I. (2004) rVISTA 2.0: Evolutionary analysis of transcription factor binding sites. *Nucleic Acids Res* **32**: W217–221.

McDonald JH, Kreitman M. (1991) Adaptive protein evolution at the *Adh* locus in *Drosophila*. *Nature* **351**: 652–654.

Ohta T. (1973) Slightly deleterious mutant substitutions in evolution. *Nature* **246**: 96–98.

Ohta T. (1993) Amino acid substitution at the *Adh* locus of Drosophila is facilitated by small population size. *Proc Natl Acad Sci USA* **90**: 4548–4551.

Schneider A, Souvorov A, Sabath N, landan G, Gonnet GH, Graur D. (2009) Estimates of positive selection are inflated by errors in sequencing, annotation, and alignment. *Genome Biol Evol* **1**: 114–118.

Schwartz S, Elnitski L, Li M *et al.* (2003) MultiPipMaker and supporting tools: Alignments and analysis of multiple genomic DNA sequences. *Nucleic Acids Res* **31**: 3518–3524.

Sunyaev S, Ramensky V, Koch I, Lathe III W, Kondrashov AS, Bork P. (2001) Prediction of deleterious human alleles. *Human Mol Genet* **10**: 591–597.

Thomas L. (1974) Biological signals for self-identificaton. In: Brent L, Holborrow J, (eds), *Progess in Imunology II*, pp. 239–347, Amsterdam, North-Holland.

Walsh EC, Sabeti P, Hutcheson HB *et al.* (2006) Searching for signals of evolutionary selection in 168 genes related to immune function. *Hum Genet* **119**: 92–102.

Yokoyama S, Tada T, Zhang H, Britt L. (2008) Elucidation of phenotypic adaptations: Molecular analyses of dim-light vision proteins in vertebrates. *Proc Natl Acad Sci USA* **105**: 13480–13485.

Zhao Z, Fu Y-X, Hewett-Emmett D, Boerwinkle E. (2003) Investigating single nucleotide polymorphism (SNP) density in the human genome and its implications for molecular evolution. *Gene* **312**: 207–213.

Index